Insights from Data with R

Insights from Data with R

An Introduction for the Life and Environmental Sciences

OWEN L. PETCHEY

Department of Evolutionary Biology and Environmental Studies,
University of Zürich, Switzerland

ANDREW P. BECKERMAN

Department of Animal and Plant Sciences, University of Sheffield, UK

NATALIE COOPER

Natural History Museum, London, UK

DYLAN Z. CHILDS

Department of Animal and Plant Sciences, University of Sheffield, UK

OXFORD
UNIVERSITY PRESS

UNIVERSITY PRESS

Great Clarendon Street, Oxford, OX2 6DP,
United Kingdom

Oxford University Press is a department of the University of Oxford.
It furthers the University's objective of excellence in research, scholarship,
and education by publishing worldwide. Oxford is a registered trade mark of
Oxford University Press in the UK and in certain other countries

First Edition published in 2021

Impression: 1

Published in the United States of America by Oxford University Press
198 Madison Avenue, New York, NY 10016, United States of America

British Library Cataloguing in Publication Data

Data available

Library of Congress Control Number: 2020948906

ISBN 978–0–19–884981–0 (hbk.)
ISBN 978–0–19–884982–7 (pbk.)

DOI: 10.1093/oso/9780198849810.001.0001

Printed in Great Britain by
Bell & Bain Ltd., Glasgow

Preface

Overview

Many activities, including experiments, surveys, clinical trials, and field-work, generate data. These data provide insights—intuitions and conclusions that come from identifying patterns in data. Insights are critical for answering questions, solving problems, guiding decisions, and formulating strategy. But getting insights from data, and doing so efficiently, reliably, and confidently, does not come easily. Yet getting insights from data is a foundational skill for all scientists.

Insights from Data with R is for life and environmental science undergraduates (though may also help anyone beginning in their learning about data analysis), and for their instructors to teach alongside. It is not about statistics per se, but about that initial transition from having collected data as part of a project to that first, and so satisfying, realization that there is a pattern in your data. It combines the elements of the successful undergraduate data analysis courses of Petchey at the University of Zürich and of Childs at the University of Sheffield, the 'Introduction to R' courses taught internationally for 15 years by all four authors, and the book *Getting Started with R: An Introduction for Biologists*, second edition, by Beckerman, Childs, and Petchey (2017), all using R with the RStudio platform.

Insights (from Data with R) first covers what insights are and why they're so important, and moves on to discuss features of data that can make it hard or easy to gain insights. It then describes how to obtain insights

Insights from data with R: An Introduction for the Life and Environmental Sciences.
Owen L. Petchey, Andrew P. Beckerman, Natalie Cooper and Dylan Z. Childs, Oxford University Press (2021). © Owen L. Petchey, Andrew P. Beckerman, Natalie Cooper and Dylan Z. Childs.
DOI: 10.1093/oso/9780198849810.001.0001

from data. Obtaining them involves knowing what you are aiming for, and then a whole lot of preparation, importing, cleaning, tidying, checking, double-checking, manipulating, and ultimately summarizing and visualizing the data.

It is common to hear people who work a lot with data say that about 80% of effort and time during real-world data analysis is spent on these kinds of tasks (and only about 20% on making statistical inference). Yet many books about data analysis ignore this 80%. They also overlook that the skills involved in this 80% are valuable in their own right. We are of the opinion that these skills alone go a long way towards allowing you to gain robust, informative insights from your data.

Insights will help you develop an efficient, reliable, and confidence-inspiring workflow for managing your data and drawing those initial insights out of them, and at the same time introduce you to core R skills for data management and visualization. Efficiency comes from learning methods of analysis that are transferable between problems and their associated datasets, and putting these methods together into an equally transferable workflow. Reliability—the ability to avoid, identify, and correct mistakes, and to reproduce work—comes from being able to evaluate multiple methods and functions and use a system of checks and balances throughout your workflow. Confidence comes from practice, encouragement, and achievement. We seek confidence that our workflows successfully generate insight.

Given our expertise and its ever-growing importance, we use R and RStudio throughout *Insights*. We use RStudio to interact with R, as it makes working with R a more pleasurable experience for the user. As in our undergraduate courses, and in the second edition of *Getting Started with R*, we teach an approach to using R based on the 'tidyverse' packages that have revolutionized data exploration and analysis in R. This approach provides a very consistent, efficient, and transferable workflow that is easily taught and learned. It is also usable with online data sources and scalable to large datasets, particularly by interfacing well with various database systems. Getting to grips with the tools to manage, summarize, and visualize small datasets like the ones we use here for insights will inspire you with confidence for much bigger ones.

Although we are biologists, and the demonstrations of getting real insights from data in *Insights* are from the biological and environmental sciences, we imagine *Insights* will be appropriate for anyone seeking to gain insights from data, and at the beginning of their journey in doing so.

The learning 'curve'

It's worth knowing what's coming. The learning curve (Figure 0.1) for this book is not a curve! It is a continual upward line, hopefully not too steep at the beginning, and never too steep, but also not so shallow that you get bored. As you work through the book you will learn more and more, while building on what came before. You should feel continually challenged (which may get a bit tiring), and perhaps at times feel a little overwhelmed, but always be clear that you, with our help, have the ability to make progress.

There will likely be some tough times, perhaps even times when you feel like you can't continue. You will be learing new vocabulary, new ways of using your computer, working with data that has problems, fixing these problems, and ultimately developing summaries and graphs to develop

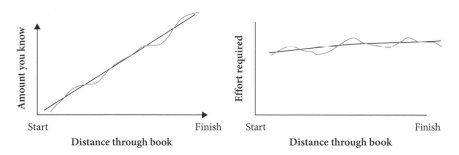

Figure 0.1 The *Insights* learning curve (left) and the effort-required curve (right). We try to make the beginning of the learning curve not steep, and then to keep you learning and learning, such that a reasonable and relatively constant effort is required.

insight. If you get stuck or frustrated, don't be afraid to take a break, have a drink and a cookie/biscuit, go for a walk, and then try again, perhaps with some help.

Untidy and dirty data

The data used in *Insights* are different from those associated with many other data analysis courses and books. The data are deliberately *disorganized*. This is quite different from many data analysis courses and books, where datasets are supplied ready for analysis. But it is also more like what you might start with from lab books, machines, or online data sources. A consequence is that the data are not visualization or analysis 'ready'. One might say the data are *untidy*. Also, the data are not provided by us; rather, you will download them from websites where the data are available to the public. Expect to spend time working with the data to get them 'research ready', getting to know the data, and learning the tricks and tips of how to do so efficiently and confidently.

No statistical tests or models

As we noted above, insights are intuitions and conclusions that come from identifying patterns in data. This does not formally require statistics. It does, however, require have developed an understanding of what the question is we (you) are trying to answer before making data summaries and graphs.

This book does not include any statistical tests, such as null hypothesis significance tests (or any other statistical tests or models), for a few reasons. (i) There is enough to be learned and gained from data analysis without such tests. We believe that the first steps in an introductory data analysis course should focus around the content of *Insights*; statistical tests can wait their turn. (ii) Statistical tests can be quite daunting and difficult, so we leave them until we have a solid hold on identifying patterns with respect to our questions that ultimately form the basis for developing appropriate statistical models and making statistical inferences. (iii) There is a risk that early learning of statistical tests encourages a rather one-dimensional view of data analysis (e.g. the dimension of a p-value), whereas in reality we need to take into account many features of the data, including why they were collected, how they were collected, and even who they were collected by. (iv) Avoiding statistics at this initial stage of data analysis forces you to focus on the questions motivating the collection of the data and expectations of patterns in the data rather than focusing on p-values and statistical significance. The great success of Hans Rosling in publicizing and explaining issues in global health and development, via brilliant and simple data visualization, is a great example of how clear messages can (sometimes) be conveyed without statistical tests.

Perhaps you are of the opinion that statistics and hypothesis testing are required for objectivity, and that without them we are just subjectively looking for patterns. If so, perhaps take a look at the article 'Many analysts, one dataset: Making transparent how variations in analytical choices affect results.'[1] There are many rather subjective choices involved in doing statistics. To be clear, we do think there is a very important, even necessary, place for statistical models and tests, but that an introduction-to-data-analysis course is not that place.

Exploratory data analysis

Exploratory data analysis (EDA) was promoted by the statistician John Tukey in his 1977 book *Exploratory Data Analysis*. The broad aim of EDA is to help us formulate and refine hypotheses that will lead to informative analyses or further data collection. The core objectives of EDA are:

- to suggest hypotheses about the causes of observed phenomena;
- to guide the selection of appropriate statistical tools and techniques;
- to assess the assumptions on which statistical analysis will be based;
- to provide a foundation for further data collection.

---

[1] https://psyarxiv.com/qkwst/

EDA involves a mix of both numerical and visual methods. Statistical methods are sometimes used to supplement EDA, but its main purpose is to facilitate understanding before diving into formal statistical modelling. Even if we think we already know what kind of analysis we need to pursue, it's always a good idea to *explore a dataset before diving into the analysis*. At the very least, this will help us to determine whether or not our plans are sensible. Very often it uncovers new patterns and insights. In a sense, this book concerns EDA. But this book is also about answering questions, including assessing the weight of evidence in support of (or against) a hypothesis. Therefore it perhaps goes a little further than EDA.

Zen and the art of 'data science'

The emergence of ever more data about ever more things, and of more and more methods, techniques, and tools for looking at these data has led to the emergence of 'data science': the science of analysing complex and large data resources. Included in data science are activities such as data collection, storage, archiving, distribution, analysis, modelling, communication, and ethics. The book *Data Science for Undergraduates: Opportunities and Options*[2] states that 'all undergraduates will benefit from a fundamental awareness of and competence in data science.' It's probably OK to think of *Insights* as a book for learning the foundations of data science, but it's also important to know that *Insights* doesn't cover lots of data science aspects (such as data archiving or ethics).

Where does Zen come into this? To gain the deepest, most robust, most interesting, most valuable insights from data we need to be 'at one with the data'. How do we achieve this heady state of mind? We need to know the *details* of the data while maintaining broad awareness of *why* we're working with the data. We must have awareness of the big picture of why we're working on the data. We need to anticipate missing values and be prepared to ask why there are missing values when one might not expect any. We need to be keen to explore the distribution of the data and perhaps ask why

---

[2] https://www.nap.edu/catalog/25104/data-science-for-undergraduates-opportunities-and-options

there are a few extreme-looking values. And we need to be OK with getting warning messages from R. Put another way, we must get stuck deeply into the details and also see the big picture. We must see every detail of every tree, and the whole forest. An article along these lines discusses how data scientists with this ability can be very competitive business consultants.[3]

Open-science trends

There is increasing movement towards making science a more open process. Part of this movement involves making data more findable, accessible, interoperable, and reusable (the FAIR guiding principles of data management and stewardship).[4] When working with the datasets in the Workflow Demonstrations in *Insights*, you might take a moment to think whether they are particularly findable, accessible, interoperable, and reusable. However, *Insights* is not about teaching you how to adhere to the FAIR guidelines—that is a story for another place, and one that is being increasingly told. *Insights* does focus on data analysis methods that are repeatable, shareable, and reliable... if there are guiding principles for data analysis, then *Insights* adheres to them!

Put another way, *Insights* teaches data analysis methods that result in high *reproducibility* (a study is reproducible if someone can take the same data and reproduce the same results as reported in the original study). Another fashion in which *Insights* assists with open science is that it teaches methods that make collaborative work rather easier than it might otherwise be, such as making our work easy for other people to understand and implement themselves, hopefully without breaking it.

Intended readers

Insights is aimed at first- or second-year undergraduates in the life and environmental sciences, to accompany their first course in 'data analysis',

---

[3] http://www.programmingr.com/content/zen-and-the-art-of-competing-against-mbas/
[4] https://www.nature.com/articles/sdata201618

and at their instructors. As far as we are aware, there is no equivalent book available (though we describe in some detail the numerous related books on the *Insights* companion website (http://insightsfromdata.io)). *Insights* purposely excludes statistical methods, so students can first master the valuable and prerequisite skills of working with data, such as manipulating, summarizing, and visualizing data. It teaches an approach to using R based on the tidyverse of add-on packages, providing efficient, reliable, and confidence-inspiring methods and workflows. Our approach to learning and teaching has developed over more than two decades and proven successful in both undergraduate courses and training programmes.

Some competencies required for beginning with this book:

- You should know your way around your computer (e.g. how to find files, make folders, install applications).
- You should know how to look at and enter data into a spreadsheet (e.g. in Excel).
- You should know how to use the internet, download files, find them on your computer, and move them to a specific folder on your computer.

How is the book organized?

Figure 0.2 shows the organization of this book, and the arrows show how you could (probably should) work through it. Nothing is very special about the organization of the first two chapters.

Chapter 1. An introduction to insights, to data, and to the demonstrations in the book and on the *Insights* companion website.[5]

Chapter 2. Getting acquainted with R and RStudio, including installing them, doing some basic calculations, and getting help.

---

[5] http://insightsfromdata.io

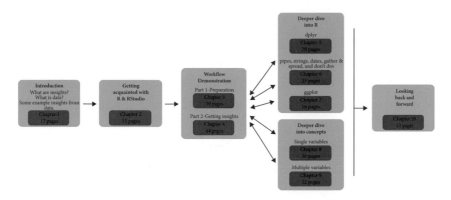

Figure 0.2 How this book is organized, and how you should work through it. This is explained in detail in the text.

Then, with *Chapters 3* and *4*, the organization of the book shifts. Chapters 3 and 4 walk through *getting insights* using an example dataset. *Chapters 5–7* contain more in-depth, complete, and detailed explanations of the mechanics of what you are doing with R and with tidyverse functions in Chapters 3 and 4. *Chapters 8 and 9* return to a focus on the example dataset and further develop core skills for insight around the various types of data in the example.

Hence, as you work through Chapters 3 and 4, you may, or may not, choose to dip into a section of Chapters 5–9. All of this is reflected in the bidirectional arrows joining Chapters 3 and 4, and Chapters 5–9 in Figure 0.2. It will be up to you how you work with these chapters; each of you is different and will probably do it differently. It will, however, likely be worth all of you being organized, for example by keeping notes about what you understood during the workflows in Chapters 3 and 4 and what you did not, and then checking this off when working through Chapters 5–9.

Here is a quick summary of Chapters 3–10.

Chapter 3 demonstrates preparation tasks, such as preparing your question, study, data, and computer, and getting data into R and ready for making insights. All of this provides a solid foundation for developing a robust workflow to gain insights from data.

Chapter 4 demonstrates getting insights, including constructing new variables, graphing data, calculating summaries (e.g. means), and evaluating patterns in the graphs and tables to gain insights.

Chapter 5 provides a deeper dive into data manipulation using tools in the **dplyr** package, including subsetting datasets, and making summaries of these subsets.

Chapter 6 provides a deeper dive into other data manipulation requirements that often arise in the life and environmental sciences. These include working with strings (words) and dates, and rearranging data from being across columns to within columns of a dataset. We also consider some formal dos and don'ts.

Chapter 7 gives an in-depth and guided explanation of how to make multiple types of graphs and enhance their capacity to provide insights using the **ggplot2** package. This builds on the introduction in Chapter 4.

Chapter 8 provides a deeper dive into evaluating features of specific variables in your data, including visualizing sample distributions and estimating numeric descriptors of central tendency (means vs medians), data dispersion, and asymmetry (variation, interquartile ranges).

Chapter 9 shifts the focus to examining patterns between two variables. The chapter includes sections on examining relationships between two numeric/continuous variables, two categorical variables (factors), and one numeric and one categorical variable. It finishes with a flurry, looking at relationships among three or more variables (including potential interactions).

Chapter 10 is the final chapter of the book, offering congratulations and some information and advice about reproducibility, an equally important subject when getting insights from data.

So, overall, you'll be learning a language of data management and visualization using R, you'll be working with example data, and you'll develop robust numerical summaries and classy visualizations of data. You certainly won't learn everything you want to know, but we can guarantee that you'll develop some excellent autonomy in learning, a platform on which to develop your *Insights from Data with R* skillset.

Online companion material

The *Insights* companion website[6] contains supplementary material including:

- an online overview of the *Insights* workflow;
- more topics in R;
- additional data analysis concepts;
- three additional Workflow Demonstrations;
- complete Workflow Demonstration R scripts;
- details of a live data analysis demonstration we often use in our introductory undergraduate classes;
- exercises and questions for each section of the book;
- more study questions and datasets that could be developed into new Workflow Demonstrations (perhaps for students to practise with and/or instructors to use);
- some related/suggested reading;

Boxes

Throughout the book are four types of box:

Efficiency and reliability. In these, we describe practices and methods for achieving higher efficiency and reliability in our journey from data to insights. They contain information about how to make our work more robust and reliable, such that it can still function if we get or add some new data, or otherwise make some changes in our work. And information to help ensure that conclusions/insights are robust.

---

[6] http://insightsfromdata.io

Be aware. These contain instructions about an opportunity/need to carefully consider an issue, for example a way to work that reduces the potential for mistakes, such as including appropriate checks and balances. These can also concern a warning or a common 'gotcha'. There are a number of common pitfalls that trip up new users of R (and more experienced users too!). We aim to highlight these and show you how to avoid them.

Action. A box containing instructions for you to do something important. Now!

Information. These aim to offer a not-too-technical discussion of how or why something works the way it does. You do not have to understand everything in these boxes to use R, but the information will help you understand how it works.

Box icon attributions:

- 'Action' by Icons Producer from the Noun Project (https://thenounproject.com/icon/1899450/).
- 'Information' by SELicon from the Noun Project (https://thenounproject.com/icon/2119887/).
- 'Efficiency and reliability' by BomSymbols from the Noun Project (https://thenounproject.com/icon/1555215/).
- 'Be aware' is the 'Warning' icon by Kristin Hogan from the Noun Project (https://thenounproject.com/icon/77514/).

All icons are licensed as Creative Commons CCBY (https://creativecommons.org/licenses/by/3.0/). Colours and sizes have been altered.

Some ideas for instructors using this book

As mentioned, we have good experiences teaching introduction-to-data-analysis undergraduate classes of 200+ students using the approach and methods in this book. Students tell us that the learning is challenging, represents a relatively high workload, is valuable, and is enjoyable. Our aim is that all students pass the course, and so far over 95% do. Here are some recommendations based on our experiences:

- The material is suitable for undergraduates with little or no prior experience of working with data, of programming, or of statistics.
- The amount of material is suitable for a six-week course of five to six hours per week (reading, practicals, and homework).
- In the first class of the first week we lead a *live data analysis demonstration*. Within one hour we go from question to answer, including collecting some data about each of the students. We believe this demonstration helps students connect with the importance and fun of the content of the course. Details of the demonstration are on the *Insights* companion website.[7]
- We have four activities each week: a lecture (sometimes in person, sometimes video lectures), before practical reading or viewing (e.g. a chapter or section of this book, or some video tutorials), a practical session (e.g. see the material at http://insightsfromdata.io), and a weekly graded assessment (administered in an automated online learning platform).
- All decisions about the course, e.g. organization, schedule, content, requirement, assessment, are taken in the context of maximizing student autonomy, purpose, and mastery, in order to nurture and stimulate students' intrinsic motivation. We treat the students like the adults they are.

---

[7] http://insightsfromdata.io

- All decisions are also taken with efficiency in mind...achieving a combination of great learning outcomes and reasonable instructor effort.
- Heavy use of automated feedback and grading leaves time for instructors to give students 1:1 support, even in a class of over 200 students.
- Owen Petchey uses the **exams** package[8] to organize a library of questions and to create examinations from these with (almost) the click of a button. The **rexams** package has options for output format, including pdf and various ones compatible with many online learning platforms.

If you have any questions about using this book as a coursebook for your undergraduate introduction-to-data-analysis course, please get in touch. We are very happy to share expertise and experiences.

Relationship with *Getting Started with R* (*GSwR*), second edition, Beckerman, Childs, and Petchey (2017)

Insights is a completely different book from *Getting Started with R*. Here are the most important differences, provided with the aim of helping you know which book to work with. If you have any uncertainty after looking at these differences, don't hesitate to contact one of us.

- What differentiates the audiences of *GSwR* and *Insights*? *GSwR*: folk who already do data work and statistics and want to learn to use R. *Insights*: folk who haven't done any data work before.
- *GSwR* motivates people who already have reasonable knowledge of getting insights from data with non-R tools to learn to use R and to implement data management, visualization, and statistical analysis with R and the tidyverse set of packages. *Insights* motivates people to

---

[8] http://www.r-exams.org

learn how to get insights from data with R and the tidyverse set of packages.

- *GSwR* assumes some prior knowledge of statistics. *Insights* assumes no prior experience of working with data or of statistics.
- *Insights* is designed as a textbook for an undergraduate 'introduction to getting knowledge from data' course. *GSwR* was not designed for this, and seems to not work very well for such purposes (though selected chapters from it combine well with chapters from other books).
- For the small amount of overlapping content, *Insights* provides more detail about how and why (rather than providing an overview tour).
- *Insights* and the *Insights* companion website[9] contain more of the content often associated with undergraduate courses than does *GSwR*, such as exercises and quizzes.

Acknowledgements

R is a product of the efforts of many individuals. RStudio is also the work of many individuals, organized by the vision of the company RStudio, whose mission is to create open source software for data analysis and statistical computing. The tidyverse collection of add-on packages was initiated by Hadley Wickham and has many contributors. We are extremely grateful to these individuals for making our data analysis and research so much more reliable, efficient, and fun. This book was written using the **bookdown** package created by Yihui Xie, which provides a suitable environment for R and RStudio users to author documents, from simple to complex.

We each have been teaching R for nearly 20 years, and in that time it is our experiences with interested, bored, critical, and all other types of student that have allowed us to become better at teaching R. We thank all the students for putting in the effort and giving their feedback about what

---

[9] http://insightsfromdata.io

INSIGHTS FROM DATA WITH R

works, about what does not, and what might work better. And we apologize to the bored and critical students for our oversights and mistakes.

We are honoured to publish with Oxford University Press and to work with its staff, particular Ian Sherman, Charles Bath, and Lucy Nash. Douglas Meekison very skilfully copyedited the manuscript. Several reviewers commented on the original book proposal, including making suggestions for improvements that were implemented.

Vanessa Mata was kind enough to place on Dryad the data used in her and her colleagues' study of bat diets. This made it possible for us to use the data and questions from the study as the basis of the Workflow Demonstration in this book.

Finally, thanks to our families for letting us have the time during evenings and holidays to work on *Insights*. We love you all, lots.

Contents

Introduction

The goal of *Insights from Data with R* (*Insights* for short) is to help you gain insights from data using highly effective tools from R, and learn valuable concepts about data analysis. After reading this book, and working along with the methods, you'll have the tools to gain insights from a variety of sources of data, using R.

1.1 What are insights?

We introduce insights here in four increasingly informal ways.

1.1.1 DICTIONARY

The Cambridge Dictionary states that insights are 'a clear, deep, and sometimes sudden understanding of a complicated problem or situation, or the ability to have such an understanding'. The Collins Dictionary states 'If someone has insight, they are able to understand complex situations.' If you search on the internet for 'what are insights', you will probably find a lot of business-related articles. And you'll soon see phrases like 'customer insight', 'actionable insights', and 'data-driven insights'.

Insights from data with R: An Introduction for the Life and Environmental Sciences. Owen L. Petchey, Andrew P. Beckerman, Natalie Cooper and Dylan Z. Childs, Oxford University Press (2021). © Owen L. Petchey, Andrew P. Beckerman, Natalie Cooper and Dylan Z. Childs. DOI: 10.1093/oso/9780198849810.003.0001

The last phrase is a good reminder to be clear that we don't think insights can only be derived from data—insights can be gained in all sorts of ways, for example by listening to and seeking to understand people.

1.1.2 THE BUSINESS PERSPECTIVE

The business perspective on insights is a useful one to know about and take inspiration from. One of the top results of your internet search for 'what are insights' might be a website called `thrivethinking`, which gives these descriptions of insights:[1]

- An unrecognized fundamental human truth.
- A new way of viewing the world that causes us to reexamine existing conventions and challenge the status quo.
- A penetrating observation about human behaviour that results in seeing consumers from a fresh perspective.
- A discovery about the underlying motivations that drive people's actions.

If thinking about the business and corporate world leaves you uninspired, or perhaps even feeling a bit nauseous, take a moment to have a look around on the internet for evidence that insights from data are being used for activities such as meeting the Sustainable Development Goals[2] (SDGs). You may find the Open SDG Data Hub,[3] which 'promotes the exploration, analysis, and use of authoritative SDG data sources for evidence-based decision-making and advocacy. Its goal is to enable data providers, managers and users to discover, understand, and communicate patterns and interrelationships in the wealth of SDG data and statistics that are now available.'

---

[1] https://thrivethinking.com/2016/03/28/what-is-insight-definition/
[2] https://www.un.org/sustainabledevelopment/sustainable-development-goals/
[3] https://unstats-undesa.opendata.arcgis.com

1.1.3 OUR DEFINITION

Insights come from identifying and recognizing patterns in data. Insights give you the opportunity and platform to either generate or evaluate hypotheses and to conjecture at the answer to questions. Perhaps the simplest framework for gaining insights is a graph or a table—these are summaries of data that are motivated by and organized around a question, objective, or expected outcome.

This is why statistical tests are not needed for insights. They are a tool to add inference to insights and to test, using various tools of probability, how likely these outcomes might be. Gaining insights comes from the simple but vital process of wrangling data to ultimately generate the graphs and tables that are motivated by your scientific quetsions.

1.1.4 OUR ECOLOGY EXAMPLE... WE LOVE FRUIT

Insights are somewhat like fruits (Figure 1.1). They can be delicious, beautiful, interesting, and useful. They might be fruits no one else knew about,

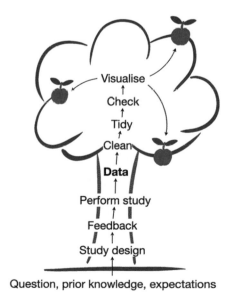

Figure 1.1 Insights sit atop, and rely upon, a foundation of well-designed and executed enquiry, the resulting data, and how those data are treated and queried.

with new colours, shapes, sugar profiles, calories, pip/pit/stone sizes, and prickly bits on the outside. Now think of what goes into making fruit: soil, seeds, water, nutrients, favourable environmental conditions, and days, months, perhaps years of growth. The DNA of the plant, movement of water and nutrients through its vessels, work (photosynthesis) driven by sunlight, and pollinators. The growth and ripening of delicious fruits requires a lot of care, work, and favourable conditions.

Getting data to bear insights is similar to growing fruits; insights, like fruits, grow from a clear and important question into a study and the resulting data, finally ripening with our help into insights.

The majority of the contents of this book are about *how to get insights from data*, i.e. how to grow a tree from a seed, to have it flower, and to produce fruit. Along the way we will make a few fruits, and we hope you find some of them delicious, beautiful, interesting, and useful. And not too prickly.

We don't use any old method to grow our tree and have it fruit. We will help you learn practices that give a *general, reliable, efficient, and robust process for getting insights from data*. Having a general, reliable, efficient, and robust process for getting insights from data should increase confidence, comfort, and enthusiasm for doing so. Furthermore, the easier the process of getting insights from data is, the more effort and focus we can devote to the insights themselves and how they affect our understanding of how our world works.

You might think we are rather overemphasizing insights from data— why have four definitions, including one based on fruit? You might think the reality is that most data-related work involves rather unglamorous data management and analysis that never result in an insight. You might be right. But our aim is that you learn with us methods that make the mundane simple, so you can spend the majority of your time thinking about what patterns in the data mean, and less time thinking about how to do the less glamorous and more tedious parts of generating the tables and graphs that deliver the capacity to make insights from data.

For example, after reading this book you will hopefully be quite repelled by the idea of moving data around in a spreadsheet to arrange them appropriately for visualization and analysis. You will certainly feel the power of making and customizing graphs and tables without using a spreadsheet. You will know that doing such work in R/RStudio is fun, flexible, powerful, efficient, reliable, and transparent.

1.2 Question, question, question (how are data born?)

We hope it is clear that gaining insight from data benefits from you having an a priori question in mind. Having a question is critical, otherwise we can get lost in our data, in a long search for patterns (and if we search long enough we will find patterns, just by chance). So a question should be your starting point; the data should not be.

Three types of question are quite common in the process of gaining insight:

- Is there a pattern/relationship that conforms to our expectations?
- Can we ascribe causation to the pattern?
- Can we predict from the pattern?

The first is about identifying patterns in a dataset with respect to expectations and distinguishing them from noise. If we find a pattern, we might have an a priori idea, or set of ideas, of why it occurs, but we probably can't say for certain just from looking at the pattern. To ascribe causation, we need, usually, to evaluate what kind of study our data come from.

Designed experiments performed in controlled conditions are the birthplace of many datasets. The number of observations, the variables they contain, and the relationships among some of the variables (e.g. the variables describing any manipulations) are determined by the researcher. In this situation there can be a very clearly specified question, hypothesis, and prediction in existence before the experiment is designed and the data are collected.

At the other end of the spectrum are datasets born from (or even adopted by) exploratory studies, where one might not really know what one is looking for, or not have a clear question, and so they record many variables, few or none of which describe experimental manipulations, and among which there is often correlation (i.e. strong association, though don't worry if you don't know what correlation is; we cover it in detail in Section 9.1).

In the middle ground are 'field' experiments (including clinical trials), in which there may be quite a few variables, a small proportion of which describe manipulations/interventions, and some of which are correlated. One might collect information about some variables because we suspect they might be important, but don't really know if they are. Such studies will progress much more smoothly, quickly, and reliably if one has a very clearly specified question, hypothesis, and prediction in existence before the experiment is designed and the data are collected.

Also somewhere in the middle ground are data from observational studies (i.e. including no manipulations/interventions) (e.g. surveys or comparative studies), which are carried out for a specific question, potentially with clear expectations/hypotheses and predictions about how groups of things respond to other variables. The data may be collected specifically for the study, or already-existing data may be used (or both). One might collect data about things that might generate patterns in the observations, and thereby accumulate numerous variables, some of which may be correlated.

While the workflows for insight acquisition for all of these sources of data can be very similar, the nature of the insights we can acquire can be constrained due to the differences in the type of study. Insights about causation are possible from designed experiments, harder from 'field' and 'observational studies', and largely not possible from purely exploratory studies.

The final type of question, *can we predict from the pattern?*, can employ a different set of tools, such as machine learning; we do not look at examples of answering this type of question in *Insights*.

Your question is thus a very important motivation for searching for insight. Researching what is already known about your question, or a related question, allows you to develop expectations about what you might find out (i.e. hypotheses), and why what you might find is worthwhile. If your question and hypothesis are directional (i.e. the direction of a pattern in the data) and quantitative (the strength of the pattern), this can make insights more focused than if your hypothesis is just that there will be some difference/pattern.

Where do our questions come from? They often come from observations of the world. If these observations are thought of as data, then one can see that questions come from data, as well as questions giving rise to data. This is a normal process: observations we make prompt a question, we answer the question by making new observations, and these new observations often raise new questions. And so on.

Being very clear about our question at the start of our study ensures we are beginning with the end in mind. A good habit to adopt is to sketch a graph that you hope to make that will clearly and quickly show the answer to your question. We also must be certain about why we are trying to answer the question, who we may want to influence, and how we will aim to do that.

1.3 But what exactly are data?

Data are records/observations/measurements of things occurring in time and space. Those things might be people, other animals, locations, or the weather, and may be collected through time and/or through space. Data can be quantitative and qualitative. The observations have values associated with them. The values of quantitative data can be continuous or only whole numbers, and may be constrained to only positive values (e.g. counts of numbers of things). Temperature, size, number of offspring, age, number of species, and elevation are examples of quantitative data. The values of qualitative data can be words that describe characteristics of the things, such as their habitat, their colour, or their name. Observations can also be dates, times, or directions (e.g. north, east, south, west).

Data usually contain variation; for example, data about the reaction times of individuals to a threat will contain some variation among the measured reaction times. It is this variation, and patterns in this variation, that we gain insights from. Hence we should always be thinking about, and aware of, the different sources of variation. Real differences among things (i.e. patterns) are important and useful sources of variation. Inaccurate and/or imprecise measurements are undesirable (though very common) sources of variation—we often call this kind of variation 'noise'. Getting insights from data involves, in considerable part, displaying variation in ways that enable us to distinguish patterns from noise.

A 'dataset' is a collection of data that is arranged conveniently for examining relationships among different properties of the things we observed. Datasets can come in many forms, but we adhere to one type of dataset: rectangular (a two-dimensional table), with rows and columns. If you've seen or used spreadsheets of data (e.g. Figure 3.2), you are probably be familiar with this type of arrangement (though spreadsheets can accomodate others also). In the dataset in Figure 3.2 each row is an observation that contains values for several variables (in the columns) about a prey item found in an individual bat's poop. The variables describing the prey items in the poop (columns) are the different things known about that prey item, e.g. which poop it was found in, features of the bat the poop came from (e.g. age, time, and date caught), and features of the prey item (e.g. taxonomy, size). We later explain more about this particular method for arranging data and why it is important.

1.4 Response and predictor variables

We use the term **response variable** for the variable that we are interested in explaining. When we make a graph, we typically put this one on the *y*-axis. When we make a summary table, we might take the mean of this variable, and report this in the table. Another term for this is the **dependent variable**—it is the variable that *depends on* or *responds to* other variables. These other variables are often termed the **predictor**, **explanatory**, or

independent variables. They usually get put on the *x*-axis of a graph, or correspond (map) to the colour, shape, or size of points or lines on a graph. In a table, they might correspond to groups, such that we might have a mean of the response variable in each group.

1.5 Some key features of datasets

To recap and expand on our discussions above, here some key features of datasets that influence and sometimes constrain our ability to draw insights from data. For a frame of reference, let's keep thinking about bats, their poop, and the insects we find in their poop.

- *Observations:* the number of things (objects) observed (e.g. number of bats).
- *Variables:* how many things about each bat were recorded (e.g. poop ID, insects in poop)?
- *Manipulation:* do any of the variables describe experimental manipulations?
- *Correlation:* are there correlations among the variables?
- *Independence:* how independent are the observations?

Number of objects observed is important because with more observations we can more reliably detect patterns. With very few data points we will often see no pattern, or a pattern that is not very strong, and if we do see a pattern it will be difficult to know if it results from chance alone. More observations give us greater ability to see real patterns, and to not perceive patterns in randomness. Lots of data is sometimes also problematic, however: one can detect very weak patterns in very large datasets (i.e. even very weak patterns can appear to have strong evidence; we can have very high confidence that a practically insignificant pattern exists). This is another good reason to forget the statistics for now and focus more on the effect size (i.e. our ability to detect a pattern for a certain number of observations) and its practical significance.

Number of variables is important because with more variables we have a greater chance of finding associations among them, even when the data are random. Take a look at the website of spurious correlations by Tyler Vigen[4] showing crazy-seeming correlations (e.g. between per capita cheese consumption and number of people who died by becoming entangled in their bedsheets). Presumably these strong correlations are found by searching among many variables. More variables also mean more combinations of variables to look at, and the whole process of getting insights becomes more challenging. A key stage in any analysis of datasets with many variables will be to try to focus on only the most important variables by thinking hard about the system being studied and the questions being asked (a couple of the Workflow Demonstrations have datasets with quite a lot of variables (30 or so), but we will reduce this number a lot, by a couple of methods).

Variables describing manipulations are important because without manipulations we only have correlations, and from them alone we cannot infer causation (or at least it is difficult to do so). That is, only if we manipulate something can we demonstrate that it has (or does not have) a causal effect on another thing. Datasets (and the studies they result from) that rely only on observation cannot demonstrate causation. Note that many scientists use the term *experiment* to describe a study even if it doesn't involve a manipulation. Two of the Workflow Demonstrations involve a study with manipulations; the other two do not. Note that we don't cover experimental design in this book, so will not explore and explain why, for example, manipulations should be randomly assigned; just know that applying a manipulation is necessary but not sufficient for appropriate design of a manipulative experiment.

Correlation among variables, and specifically among the variables we are using to do the explaining, is generally problematic, but can be quite common. Say we want to know what about people explains differences in their IQ, and so we record things about them such as time spent studying, the IQ of their parents, time spent in unsupervised play, socio-economic

---

[4] https://www.tylervigen.com/spurious-correlations

status, and so on. Many of these characteristics will be quite strongly associated. When this occurs, we can't easily tell which variables are more important explanations than others, because they explain similar variation in the data (IQ here).

Independence of the observations is important because this, as well as the number of observations, determines how likely we are to see a pattern that exists, or to not see a pattern that doesn't exist. There are many ways by which observations can be non-independent… two observations might come from the same individual, or from the same mother, or the same location, or the same time. Anything that observations share in common can create non-independence. Each of the Workflow Demonstrations contains some non-independence among observations; we will discuss this and then show how to deal with it.

In summary, if we have lots of observed objects, low correlation among variables, few variables, independent observations, and manipulations, we can expect to sail through the journey from data to insights. If we have some combination of few observations, many strongly correlated variables, non-independent observations, and no manipulations, we are going to have to work a bit harder to get the insights, or perhaps even have to see if we can find a way to redesign our study and data collection to avoid what could be a very difficult data analysis problem.

1.6 Demonstrations of getting insights from data

This book contains one core demonstration of getting insights from data. The *Insights* companion website (http://insightsfromdata.io) contains three more demonstrations. Our primary objective in the book is to deliver to you a single dataset that contains the opportunity for substantive insights, but also contains many issues that get in the way of obtaining insights. We want to show you how to fix these issues and the techniques that you can use to acquire deep and meaningful insights.

The examples were chosen for their set of issues and for their diversity and, hopefully, intrinsic interest. All of them are about food. Why food? No particular reason, other than we all likely have some interest in food, getting enough, not getting too much. Food is also a matter of global concern—the second Sustainable Development Goal is 'Zero Hunger'. So we will look at questions, puzzles, and problems concerning food.

The data span a wide range of questions and approaches, from tightly controlled laboratory experiments with manipulations to an observational study of global patterns. With such a range of case studies you will experience a wide range of methods for getting insights from data, as well as understanding the strengths and weaknesses of these different approaches and the data in them.

In the following text we give brief descriptions of each of the four demonstrations and a sketch of the graph we will use to answer the question in each. Each graph contains a possible outcome of the study, and we show a hand-drawn sketch because this is what we advise you to make at an early stage in such a study. We also mention the insight that would be revealed by such an outcome.

Demonstration data 1: Do male and female bats have different diets? And does any sex difference depend on the age of the bat? This is the study we work through in the book. It is a study of the diet of over a hundred wild bats. The expectation was that the previously observed body size difference between male and female bats (females being larger) would translate into a diet difference. Diets (they eat insects) were observed and recorded from molecular analyses of each bat's poop. There was no manipulation/intervention, there are relatively few explanatory variables, and not a great deal of correlation among them. Difference in number of prey in diet, average size of prey, and proportion of migratory prey items were examined. The question of whether any sex differences are dependent on age is a question about an *interaction*. The study is real and was published in 2016; you will pretend you are the researcher while working through the demonstration, to attempt to give you an impression of the thrill of getting insights. You will get the data from an online archive of the real data that the researchers gathered and analysed. A possible outcome from the analysis of these data is sketched in Figure 1.2. In the case of this hypothetical result, the insight

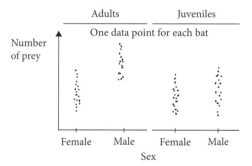

Figure 1.2 A hypothetical outcome of the demonstration study of sex and age differences in three features of bat diets.

would be that while adult males have more diverse diets than adult females, there is no such sex difference observed in juvenile bats. That is, the sex difference in diet diversity depends on the age of the bats; it only happens in the adults. Furthermore, even among adults, some individual female bats have more diverse diets than some males, i.e. there is considerable variation around the observed trend.

Demonstration data 2 (online): Does the political system of a country correlate with the diversity of food available? We hypothesize that countries with democratic political systems will have more diversity of food available than countries with autocratic systems. We will restrict ourselves to two variables, diversity of food available and political system, thus making the problem simple, but also rather limited in what one can conclude (if anything) about causation. Inference about causation is further limited by the lack of any manipulations/interventions, i.e. this is a purely obser-vational study. As far as we know, the analyses are novel. Any insights will be new! Though, since the findings are not published/peer reviewed, they should be treated as preliminary and (as usual) be critically assessed. There are a couple of hundred data points (one for each country of the world), and no correlation among explanatory variables, because we are using only one. There is non-independence in the data, and this will be discussed. A possible outcome of the study is sketched in Figure 1.3. In the case of this hypothetical result, the insight would be that countries

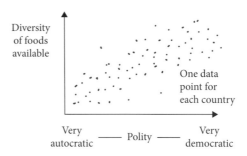

Figure 1.3 A hypothetical outcome of the demonstration study of the relationship between the political system of a country and the diversity of food available.

with more democratic political systems (polities) have more diverse foods available, though with considerable variation around this general trend.

Demonstration data 3 (online): Does a more diverse human diet lead to more stable populations? The study was designed to address the hypotheses that more pathways of energy flow to a consumer (i.e. more prey species) stabilize the consumer population dynamics, (1) by increasing the abundance of the predator, (2) by reducing variability/fluctuations, and (3) by delaying extinction. It involves a manipulation: the diversity of prey available is an experimental manipulation, and therefore insights can be obtained about causation. The analyses are relatively straightforward, but this demonstration has some complexity in the calculations we must do to obtain the response variables (i.e. the variables we examine to answer our biological question). There are relatively few data points, and the study was done in very controlled laboratory conditions. The study was real and was published in 2000 by Owen Petchey. You will get from an online archive a copy of the raw data, which contain non-independence that will be discussed. A possible outcome of the study is sketched in Figure 1.4. In the case of this hypothetical result, the insight would be of a quite strong effect of prey diversity on consumer population abundance (for example). The effect is considered strong because there is relatively little variation among the different compositions within the same prey diversity level, compared to the differences among prey diversity levels.

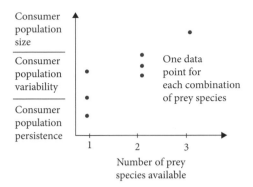

Figure 1.4 A hypothetical outcome of the demonstration study of how a manipulation of diet diversity affects any one of three features of the consumer population.

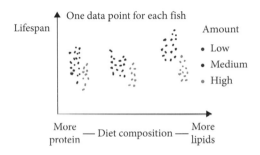

Figure 1.5 A hypothetical outcome of the demonstration study of how a manipulation of diet composition and amount affects the lifespan of fish.

Demonstration data 4 (online): How do diet composition and amount of food individually, and in combination, affect individual characteristics related to health and fitness? Recent research suggests that, in some situations at least, calorific restriction (eating fewer calories) can have positive health effects, such as reduced mortality and other health benefits. This study compared the effects of amount of food with composition of food, by manipulating the diets of small fish in controlled laboratory conditions. There were several hundred fish and quite a number of explanatory variables, including those describing the manipulation. The study is real and was published in 2018. You will get from an online archive a copy of the real data gathered and analysed. A possible outcome of the study is sketched in Figure 1.5. In the case of this hypothetical result, the insight would be that there is a considerable variation in lifespan regardless of the experimental treatment (diet composition or amount of food available). There is some evidence of shorter lifespan at high amounts of food, and longer lifespan with more lipids, though only in the low- and medium-food-amount treatments. These results lead to insights of some contingency in the effects of diet composition and amount. It is not as simple as one or the other being more or less important. Rather, the effect of one depends on the other. Note that this is presented not as a demonstration, but rather as a series of questions that guide you through the workflow (solutions are also provided).

1.7 The general *Insights* workflow

We believe that repeatedly using the same systematic workflow can help us make sure we do everything we need to. And revising it appropriately can help us make it even better. We propose a relatively consistent workflow for getting insights from data. The description of it in Figure 1.6 has 21 steps,

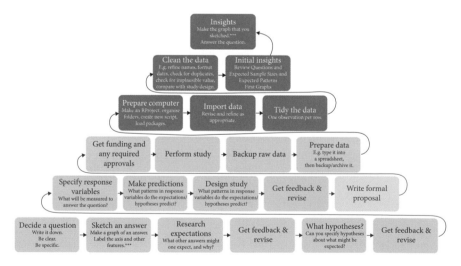

Figure 1.6 A common workflow for gaining insights from data. Work from the bottom left, along the row from left to right, and gradually up the levels. The asterisks *** emphasize the link between the graphs sketched at the beginning and the end of the workflow.

which may seem like a lot, but even some of these could be broken up into multiple steps. In particular, the two steps that involve insights at the top of the pile could be broken up quite a lot more, and perhaps be repeated several times for different questions/insights from the same study.

You may also notice that the first 12 steps (the first two and a half layers in Figure 1.6) are done before you perform the study, and are the absolutely essential foundation of what then comes. We will again and again emphasize elements of this preparatory phase, including the importance of a clear and specific question, of sketching a graph that could answer it, doing research about what is already known (building expectations), specifying what will be measured and what will be our response variable, and the very important process of getting feedback and improving the preparation/planning/proposal.

The eight steps in the upper two and a half layers of Figure 1.6 are about going from the data to the insights, i.e. the answers to our questions. We will spend quite a lot of time learning the details of these. In fact, we expand these here for repeated reference throughout the book (Figure 1.7).

There is a final and essential step that we have not included in this figure: communicating our insights. Science communication is such an important, extensive, and nuanced subject that we will not attempt to consider it here. But you must. If we don't effectively communicate our findings, we might as well have not done the work.

This workflow is available as a checklist on the *Insights* companion website.[5]

1.8 Summing up and looking forward

You now have an idea of what we mean by insights and the different types of questions and study types that can affect our ability to gain insight from data. We have also introduced key features of data that further influence our ability to acquire insights, including the number of objects observed,

---

[5] http://insightsfromdata.io

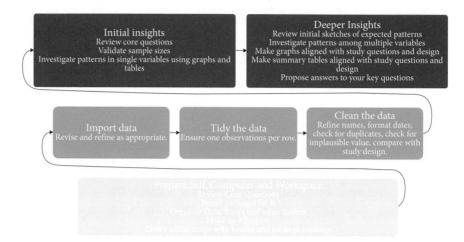

Figure 1.7 The portion of the workflow we focus on in this book, from preparing yourself and computer to work with data, to gaining advanced insights from your data.

the number of variables recorded about each observed object, if there were manipulations, the amount of correlations among variables, and the amount and nature of non-independence among the observed objects. And, finally, we have introduced four demonstration datasets and the concept of gaining an insight from these data. This book contains an in-depth work-through of the bat diet study.

Now, before we dive into the bat diet Workflow Demonstration (in Chapter 3), let's first get you up to speed and acquainted with some of the tools we will use and, in particular, have a gentle and friendly introduction to setting up your computer to use R and RStudio. You can see we are starting at the bottom of our expanded *Insights* workflow diagram in Figure 1.7. Let's get started.

Getting acquainted

2.1 Getting acquainted with R and RStudio

Before we start working through a real example of getting insights from data, we need to become acquainted with some tools. It's a bit like learning about the feel and texture of a paintbrush, and the properties of the paints and the canvas, and the dynamics of how the paint moves from paintbrush to canvas. And, yes, insights are like art and fruit.

Learning a bit about the tools first can help us feel more confident and comfortable when we then come to use them for real. Hence we will take some time to introduce you to R and RStudio, including how to set up your computer to make data import easy, writing some simple commands, writing scripts, and what to do when things seem to go wrong (as they often do!). We'll also introduce the idea of functions, how to get and use add-on packages, how to get help, and some common pitfalls to avoid.

If you're already familiar with R, great! You'll whizz through this material, but please don't ignore it. It's likely that our approach is a bit different from your own and can offer some refinement that reduces the potential for errors and frustration along the way. Particularly, our use of projects via RStudio. Furthermore, the remainder of the book relies on knowing these initial stages.

Insights from data with R: An Introduction for the Life and Environmental Sciences. Owen L. Petchey, Andrew P. Beckerman, Natalie Cooper and Dylan Z. Childs, Oxford University Press (2021). © Owen L. Petchey, Andrew P. Beckerman, Natalie Cooper and Dylan Z. Childs. DOI: 10.1093/oso/9780198849810.003.0002

This is quite a long chapter! It has eight sections. We suggest that you
don't try to go through them in one sitting. Rather, make a plan now
of how you will tackle them. For example, 'Today I will work through
four sections, and between each I will have a 30 minute break. I will
make notes about what I learn and what I don't understand. Then I will
celebrate my progress with exercise/chocolate/beer/cooking/yoga (or
all five!). If I don't get through four sections, but I gave it a great try,
then I will also celebrate (you still deserve those five prizes!). If I get
distracted, but realize that and try to refocus, I will also celebrate (still
5/5 for energy!).' It really is worth making such a plan. Not just for this
chapter, but for each chapter when you start it. Take a look at the table
of contents to see how much, and what, is coming, and make a plan
that involves you being kind to yourself.

2.1.1 WHY R?

In the late 1990s, two academics, Ross Ihaka and Robert Gentleman, at
the University of Auckland, decided to create an open source language to
help researchers in computational statistics. Based on the 'S' language by
John Chambers, R was born—and now R is both the language and the
application that runs it.

R allows us to do with one application things that used to require several.
Data manipulation, visualization, statistics, simulations, and much more,
all in the same application. Some of us would, albeit many years ago, use a
combination of Excel, SAS, and Sigmaplot to quite successfully do what we
now do with only R. It worked, but it was not pretty or efficient.

Another great feature of R is that we write scripts. These are reusable,
readable, shareable, and beautiful instructions that should be a complete
record of our analyses. This means that if we acquire some more data, or
alter our analysis, we can repeat it with little effort, since all the instructions
for how to do it are written out and saved.

Yet another nice feature of R is that it is open source and cross-platform
(e.g. runs on Windows, Mac, and Linux). This makes it more likely that

we can keep using R/RStudio ourselves, and that our team members can. Furthermore, we can, if we like, use RStudio via a cloud service in a browser (e.g. RStudio Cloud[1]).

For when R seems to not be behaving so well, R has a very large user community, growing particularly quickly in the last ten or so years. In the early 2000s there were few books on R, and help was mostly from the help files (we will use these a bit, but they can be rather opaque) and rather daunting R mailing lists. Since then, hundreds of books, online tutorials, and online courses have been helping people learn R. We now have a vast R ecosystem, working with, developing, and supporting people using R in business and academia. When we encounter a problem in R, or wonder how we can do something, chances are someone has solved it, done it, and written about it online, or made a video showing how it's done. Later in this chapter we'll show you numerous ways to get help.

As well as being great for working with data, R is a quite comprehensive programming language. Should you desire, you can write books and reports, construct numerical simulation models, do matrix maths, get information from websites, and even send emails from within R! (None of this is covered in this book.)

All that being so, it's often said that the initial learning curve is very steep. We start our ascent, and could get rather dizzy, get put off, and lose confidence and motivation. This is where we come in, providing an efficient, reliable, and confidence-inspiring approach to learning and using R.

2.1.2 WHY RSTUDIO?

RStudio is a helper application—it sits between you and the R application. It's a bit like a window through which you look at and interact with R. There are quite a few such helper applications for R, but RStudio is our preferred choice. Reasons include that it does the basics really well—i.e. it allows us to interact with R easily and simply. It doesn't have a lot of overheads. It makes lots of things quite a bit easier or just nicer. (And you can really

---

[1] https://rstudio.cloud

easily change the colours of your windows... always important to have colours you like!) Another thing RStudio has is *projects*... a very convenient tool for keeping us organized and making switching between projects quite painless. We will describe projects in RStudio below, and then use them widely throughout this book.

Just like R, RStudio is cross-platform. Even better, its appearance is exactly the same on different computers and platforms (well, there are things you can change, like the colours, but they don't change how it works!). Learning to use R via RStudio on a Windows machine will look the same as on a Mac or Linux machine. This is a big advantage if you think you might ever use more than one type of computer, or are teaching students who use their own and therefore probably quite diverse hardware.

2.1.3 GETTING AND INSTALLING R

Download R from the Comprehensive R Archive Network (CRAN[2]). Get the 'base distribution'. Get the distribution appropriate for your computer, i.e. for Windows[3] or for a Mac[4] (be sure to get the version for your OS X version). Install like you would any other application. Don't mess with anything... the default settings will be just fine, and changing them could lead to problems. We do not recommend adding an icon to the desktop when Windows asks. That's a no.

    Download and install R now!

After installing R, it should be visible in the Programs menu on a Windows computer or in the Applications folder on a Mac. However, it would be a good idea to read the next section before launching R...

---

[2] http://cran.r-project.org
[3] http://cran.r-project.org/bin/windows/base/
[4] http://cran.r-project.org/bin/macosx/

2.1.4 GETTING AND INSTALLING RSTUDIO

First, be very clear in your mind that RStudio is a completely different program from R. R will do the work for us, and RStudio will be your helper/buddy/friend in using R. You must download and install RStudio separately from R.

RStudio was created by and is developed and maintained by RStudio (it's also the name of the company). The company is for profit and they sell software tools and services to big and small businesses, making many products freely available to academia. But even though RStudio is a for-profit company, they made RStudio (the fully functioning desktop version) *open source.* This is really important because it means anyone can have the source code, and therefore gives confidence that we can continue to get and use RStudio for some reasonably long time into the future. We truly believe that RStudio's balance of contributing to the R and science communities and still profiting from the tools and services it provides is quite exceptional. We trust them. So can you.

Download RStudio from the RStudio download page.[5] Get the Open Source Edition of RStudio Desktop. Install it into your Applications or Programs folder like you would any other application.

Don't get the commercial version. Don't get RStudio Server. Just get the free RStudio Desktop.

Download and install RStudio Desktop, Open Source Edition, now!

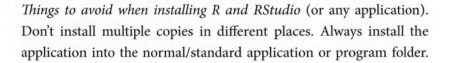

Things to avoid when installing R and RStudio (or any application). Don't install multiple copies in different places. Always install the application into the normal/standard application or program folder.

---

[5] http://www.rstudio.com/products/RStudio/#Desk

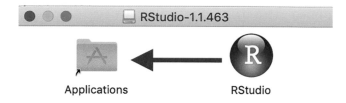

Drag and drop the RStudio icon into the Applications folder

Figure 2.1 When installing RStudio on a Mac be sure to drag the RStudio icon into the Applications folder, then close this window and delete the '.img' file downloaded from RStudio.

On a Mac, be sure to drag the RStudio file into the Applications folder shortcut (e.g. Figure 2.1). Don't try and do anything *special* unless you really know you should be doing so.

2.1.5 A BRIEF TOUR OF RSTUDIO

We suggest using R via RStudio, so let's see how that works. Open RStudio. If it's the first time, you will see something very similar to what's in Figure 2.2. You'll likely see three panes inside a single window.

1. The **Console**. This is a bit like the engine of R. We put commands in the Console, and get information back about what happened. We can type commands directly into the Console, or send them there from other places. All this happens at the prompt, >.
2. A pane with two tabs. The first, *Environment*, tab shows the objects in R's memory... for example datasets we have imported. Some buttons help us get data into and out of R. The second, *History*, tab shows the commands we previously sent to R. Some buttons allow us to reuse and keep these commands. You might see another pane here (*Connections*) that you can ignore—we won't be looking at it.

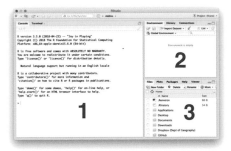

Figure 2.2 RStudio when you first open it, with the three panes numbered by us. See the text for details about each pane.

3. OMG! A pane with five tabs!
   - *Files* allows us to see, explore, and do stuff to files on our computer.
   - *Plots* shows any figures we make, and gives some buttons for working with these (e.g. exporting them).
   - *Packages* helps us manage the add-on packages we install to extend the functionality of R.
   - *Help* gives access to various help pages.
   - *Viewer* can be used to view HTML documents made in R—we won't be using it in this book.

No problem if RStudio looks a little different for you. You might even like to make it look different: take a quick look at the Tools > Global Options… menu to see your options! You might have a fourth pane—the one where you write your R scripts. RStudio tries to open looking like it did when you closed it, so if you've already messed with RStudio's appearance, or left a script open last time you closed it, you will see these things when you open it. No worries!

While the core of RStudio remains the same, RStudio is continually being developed and updated, with new features appearing and some old ones being improved. Such improvements are generally a good thing, and are done carefully so as to only very rarely cause problems for your existing code. See Section 2.6.5 for advice about updating R, RStudio, and add-on packages.

Good. We now have R and RStudio and are ready to have a play—to do some calculations in R/RStudio!

2.2 Your first R command!

Let's start simple: open RStudio, then click your mouse anywhere in the Console pane and type 1 + 3 (it doesn't matter whether you put spaces in there or not). It should appear at the prompt >. Then press the Enter key on your keyboard:

```
1 + 3
```

```
## [1] 4
```

The first line above shows the request we gave R. The next line, beginning with ##, is R's response (you will not see the ## in your Console—they are printed in this book to indicate R output). We asked R to add 1 and 3, and it responded with 4. Great!

In case you wondered, the [1] in the output is R telling us that 4 is the *first element* of the answer to our question. There is only one element in the answer to this particular question, so there is only [1]. (If there were more elements we might also see [2], [3], [4], and more.)

Throughout this book you will see that different parts of our R commands have different colours (sometimes known as *syntax highlighting)*. We did not do this ourselves… it is done automatically to help us read the code. We could explain all the different parts and different colours here. But then we would have to explain the grammar and syntax in more detail than we feel is currently useful. So we don't.

2.2.1 GETTING TO KNOW R A LITTLE BETTER

One of the best ways to get to know R is to play. It's pretty difficult to break R, so don't worry about that. Type whatever you like into the Console, press enter, and see what happens… probably just an error message; no smoke should come from your computer!

Try some simple arithmetic. Add some numbers together, subtract some others:

```
1 + 7
```

```
## [1] 8
```

```
13 - 10
```

```
## [1] 3
```

Multiply and divide some:

```
4 * 6
```

```
## [1] 24
```

```
12 / 3
```

```
## [1] 4
```

Raise one number to the power of another:

```
3 ^ 5
```

```
## [1] 243
```

Do a combination of these, noting that the order in which R performs the calculations is the standard one: (1) exponents and roots, (2) multiplication and division, (3) addition and subtraction. If you want to control the order, you can use parentheses to surround the parts you would like done first. Notice the different outcomes of these two commands:

```
3 ^ 2 - 5 / 2
```

```
## [1] 6.5
```

```
(3 ^ 2 - 5) / 2
```

```
## [1] 2
```

Similarly, if we want to raise a number to a fraction, we need to surround the fraction with parentheses. Notice the different outcomes of these two commands:

```
16 ^ 1 / 2
```

```
## [1] 8
```

```
16 ^ (1 / 2)
```

```
## [1] 4
```

The first one calculates 16 raised to the power 1 (16) and then divides the answer by two. The second one raises 16 to the power of a half. Big difference!

While the cursor is in the Console, you can press the up arrow on your keyboard to get your previous commands, and then run them again, or you can edit them using the left and right arrows to move around, then deleting text and writing in new text. We'll very soon look at another nice, even essential, way to reuse commands (saving them in script files).

2.2.2 STORING AND REUSING RESULTS

With the simple and short questions above, we are happy to just see the answer. But our questions are often more complex, taking a number of different steps, and we benefit from being able to store the answer to one part of the puzzle and use it in the next part. The solution is very simple: we *assign* the answer to a name:

```
a <- 1 + 2
```

This literally says 'please assign the answer of 1 + 2 to the name a'. It uses the **assignment operator**, written as a left arrow < -, to make the assignment.

Now we can look at what a is by typing it in the Console and pressing the Enter key on your keyboard:

```
a
```

```
## [1] 3
```

And we can use it in further calculations, e.g.

```
2 * a
```

```
## [1] 6
```

We'll use the assignment operator < - *a lot*. Typing < and then - over and over again is tiring. Use the RStudio shortcut. Click in the Console, hold down the Alt key on your keyboard (or the 'Option' key on a Mac keyboard), and press the - key. A < - magically appears!

You can, if you like, use the equals sign = instead of the assignment arrow < -. We always use the assignment arrow. It is good to be consistent.

What happens when we assign a value to an already existing name, for example by typing a < - 10? Well, the previous assignment is lost and replaced with the new one:

```
a <- 10
a
```

```
## [1] 10
```

With the assignment operator, we can assign more or less anything to a name. Here we assign to a new name b the value of 2 * a:

```
b <- 2 * a
```

Here we make c equal to b (i.e. we copy b into c):

```
c <- b
```

Note that if we now change the value of b then the value of c is unaltered:

```
b <- 2.5
c
```

```
## [1] 20
```

The objects b and c are totally independent of each other, in the sense that changing one does not change the other. This is really good, because it means we can only change objects by explicitly making them change. R won't change things without you telling it to.

Remember the Environment tab of the RStudio pane number 2 in Figure 2.2)? You should now see that R's memory environment is no longer empty. It should contain three objects a, b, and c and show their values. This information is often very useful for quickly seeing what objects you have in R's memory, and what they contain.

RStudio will, by default, save the objects in its memory when you close it, and then restore them when you next open RStudio. It might seem nice to be able to close down R, reopen it, and pick up where you left off, but it's actually rather dangerous. The details of why are a bit messy, but trust us, *don't do this*. Assume that when you close R, everything in its memory is lost. (We show below that by keeping the commands you used, you can easily recreate what was in R's memory.) To stop RStudio from saving objects by default, go to the Preferences options 'Save workspace to .RData on exit:' and change the selection to 'Never'.

2.2.3 WHAT NAMES SHOULD I USE?

So far we have used single letters, a, b, c. We could use the words gwen, tom, and nancy instead, as they are **legal** names as far as R is concerned. Legal names in R must begin with a letter, which can be followed by any sequence of letters, numbers, ., or _. Upper- and lower-case letters are allowed. For example, var_1, var.1, var1, VAR1, and myVar1 are all legal names. But 1var and _var1 are not—they don't begin with a letter. Try to remember that R is case sensitive, i.e. My_var and my_var are different. We often forget this, or how we've capitalized a name; probably you will too. While it's possible to make a name that contains a space, it's not such a good idea, because they are a little tricky to use.

- Use names that you can't easily confuse. That is, using both `var` and `Var` at the same time, with them meaning different things, is, well, unwise. They could very easily be mixed up, with disastrous effects that could be very difficult to see.
- Don't put a `.` at the beginning of a name, as then it will be hidden (unless you really really know you want to hide it!).

Shorter names are faster to type. Intuitive and meaningful variable names are easier to remember. Variable names that don't contain spaces are easier to work with.

2.3 Writing scripts

In the previous section, we typed commands directly into the Console. If we want to run a series of separate commands, and perhaps run them again in the future, or alter one of the commands, thus needing to rerun all the subsequent ones, then typing commands in the Console is not the best. We can use the up arrow to get previous commands, but there is a better solution: put your commands in a **script**. We'll review this again, but a script is a *repeatable, shareable, annotated, and cross-platform record of what you do with data to gain insight*. That's cool (we think so).

For anything but the very simplest task, put your commands in a script.

To create a new script in RStudio, navigate to the File > New File > R Script menu. You will see the new script file open in a new, fourth pane with a

Figure 2.3 RStudio when you open a new script file. The new pane with the red star is where we type our commands and other things.

tab named Untitled1 (Figure 2.3). In your new script, type some of the basic arithmetic and assignment commands we used in the previous section. The first thing you might notice is bits of your commands are automatically shown in different colours (though perhaps not the same as the colours here)—this is to help us read the script (R does not care).

When we write instructions for R in a script, we should make sure all our instructions are there, and *do not type any directly into the Console*. If we type some in the script and some in the Console, we will soon have a big mess. For example, the script might seem to work, but then magically start not to work at some point in the future, creating great frustration.

2.3.1 COMMENTS IN YOUR SCRIPTS

Annotating your instructions (providing yourself and others with some 'insight' into why you are doing things) is a vital aspect of a robust workflow for gaining insights from data. In a week, or a month, or a year, when you come back to a script you were working on, you will often wonder what some command is for, why you did something the way you did it, why you did it, and why you didn't do something else. While it's possible to work these out from the commands themselves, it is very, very, very useful to make notes to your future self. The same goes if you think someone else might ever read and work with your script. Make helpful notes for them. These are not commands for R; they are notes/comments for only humans to read.

In R, we make comments by starting one with a # (which is variously known as the hash, hashtag, number, or pound symbol). Everything in that line after the # is then ignored by R. The comments are only for us humans.

Put the following comment in your script:

```
# I really love R!
```

Some people say 'always comment your script', but what does this mean exactly? How much commenting is enough? Should we have a comment for every line of R commands? Or more than this, or less? In reality, commenting our script can easily seem a bit of a waste of time, a bit tedious, and then get skipped. The key is to write useful comments, and not write useless ones. Knowing what is useful and useless takes a bit of practice and experience. For example, definitely write a comment about a non-obvious but important choice you made. Perhaps even write a comment like # Import the data when the next line of R command clearly does this (we do this, just to be sure!).

 While it takes a little time to write comments, doing so will save time in the future. We promise. Take our word for it. Try to think of what will help you understand the script in a year's time, and then make that comment.

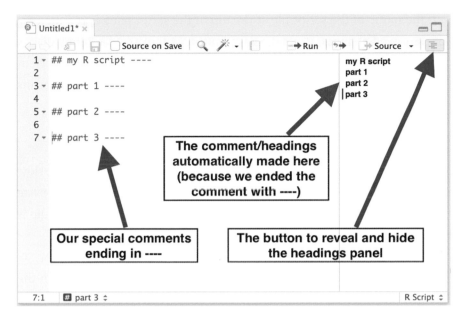

Figure 2.4 The script editor pane with special comment headings, the associated panel of headings, and the button to reveal/hide this panel.

If at the end of a comment line you write - - - -, you will see a little downwards triangle appear by the line number in the left margin of the script editor pane. If you click this, you hide the code until the next comment ending in - - - -. Also, these special comments become headings that you can use to quickly navigate through your code. See Figure 2.4.

2.3.2 SAVE AND KEEP SAFE YOUR SCRIPT FILE

Our script contains all the instructions for and details of our journey from data to insights. We need to save it. We need to keep it safe. We need to be able to find it. Alongside the data, our script is the most valuable thing in our R-life. With those, we can recreate all our insights. We don't need to save anything else—it's probably better that we don't save anything else (keep it simple).

If the name of the script file at the top of the script pane is red then we have unsaved changes. Press the Save button (it looks like a floppy disk[6]), or go to the menu File > Save As..., or use the keyboard shortcut `Control+S`. Give the file a sensible name, a name that means something to us.

2.3.3 RUNNING YOUR SCRIPTS

To run a command in our script, we need to get it into the Console. The slow and tedious way to do this is to select some of the script, copy it, and paste it into the Console. But we want a fast and efficient method, and RStudio is happy to help with a few options:

- Top right of the script pane is a `Run` button. Pressing it will run the line of code that the cursor is in if no code is selected, or any code that is selected. If it runs a line, then the cursor automatically moves to the next line, so you can run a series of consecutive lines of commands by repeatedly pressing the `Run` button.
- Clicking with the mouse is also a bit slow, so there is a keyboard shortcut. Pressing `Control+Enter` (or `Command+Enter` on a Mac) is the same as pressing the `Run` button.
- If you want to run a line of code, click anywhere in that line, and then `Run`.
- If you want to run all the commands, press Control+A (or Command+A on a Mac) to highlight all the code, then click `Run`, or hit `Control+Enter/Command+Enter`.

2.4 When things go wrong...

...and they do and will. R is *very* pedantic. Even the smallest typo can result in failure, and typos are almost impossible to avoid completely. So we must accept that we are going to make mistakes, things will go wrong, we are not

---

[6] https://en.wikipedia.org/wiki/Floppy_disk

perfect. The approach to R that we will be teaching in the next few chapters is designed to reduce the chance of making mistakes.

One type of mistake causes an error and the code fails to run. The other type of error is when the code runs just fine, but doesn't do what we intended. The first kind of mistake is so easy to make that most people, even us R-experts, do it quite regularly. Often these mistakes are caused by typos, missing commas, or missing brackets. They're nothing to be embarrassed, worried, or surprised about. We just need to be able to correct them.

The other type of mistake is more insidious—our code will run, R won't complain, but the results are garbage. Adding together the wrong two numbers would be such a mistake. Or adding numbers rather than multiplying them. We need to do everything we can to avoid making such mistakes, and to identify them when they happen. A good attitude to have is 'my code is probably not doing what I intend' and to have to prove to oneself that it is (for example by checking the result a different way, e.g. by hand, at least for relatively simple calculations).

2.4.1 ERRORS

You have probably already experienced some error messages, so we should talk about them before going any further. Let's make an error:

```
z + 10
```

```
Error: object 'z' not found
```

R tells us there is an error. It couldn't find the object z. That is, we asked R to add 10 to z, but R didn't have anything in its memory named z.

Error messages like this, and ones where the message is much longer and more complex, are scary! People see the word 'Error' and run away from their computers. Or at least have a little panic, which often involves not reading the error message. Or get instantly frustrated they got *another error message*!

Try to train yourself to respond differently to error messages. Perhaps, in your head, call them 'help messages'. Or do a little meditation when you see one. Or have a bowl of (small) chocolates on your desk, and 'reward' yourself with one every time you get an error message. Most of all, *always read the error message*. With time and experience they will become more and more useful (though sometimes we, with all our years of experience, get ones that baffle us, at which point we seek help).

2.4.2 WARNINGS

Everything we just said about error messages also applies to warning messages. The difference is that your code will still run if you get a warning message, but it might not be doing what you intended, or R has recognized that something you asked it to do is a bit risky. A good rule for warning messages is *always read the warning message* and *always understand what the warning message means*, and *always try to do something to get rid of it. Never ignore a warning message.* You may figure out that you can safely use a script that gives a warning message, in which case it is a great idea to write your future self and other people a comment about how you figured this out and why it's OK to ignore the warning.

2.4.3 THE DREADED +

Another type of mistake is one that results in the Console looking something like this:

```
> (10 + 5 + 3
+ |
```

Create it yourself by typing into the Console (10 + 5 * 3 and pressing Enter.

This is not a complete R command—it's missing a closing parenthesis,). R's response to this is to wait for us to finish the command, here by closing the parenthesis. The + at the Console is R's way of saying 'I'm waiting for you to finish, please'. When you see the +, hit the Escape key to get back to the prompt > and then Run the whole command, perhaps after completing it in the script.

2.5 Functions

Functions are the tools in our R toolbox. R comes with numerous ready-made functions for us to use. Each one helps us do some data-related tasks. Take for example the function that we can use to round a number to a certain number of significant digits—no surprises this function is called round. Here's an example of how we use it (don't run it yet):

```
round(x = 2.4326782647, digits = 2)
```

We start the command with the function name (round in this case). The name is followed by an opening parenthesis. Then come the *arguments*, each of which is separated by a comma. Then comes a closing parenthesis.

What are arguments? In the example above, there are two arguments. The first is x = 2.4326782647 and the second is digits = 2. So arguments are the information we give the function. (Probably the internet knows why this information is termed 'arguments'—feel free to look it up if you're curious).

One more thing about how we used round above: each of the arguments is of the form name = value (this form is often called a 'name–value pair' for obvious reasons). The name tells the function which argument we are giving the value for. That is, the second argument, digits = 2, says the value of the digits argument is 2—please round the number to two significant digits, thanks! (The first argument, x = 2.4326782647, is the number we would like to be rounded.)

> When giving the value of an argument, always use an =. *Do not* use the assignment operator < - inside the parentheses when working with functions. This is a 'trust us' situation: you don't need to know why, you don't want to know why, and we're not telling you ☺

How do we know what the names of the arguments are for a particular function? For example, how do we know digits is the name of the argument of the function round? The answer is that we have to look at the instructions for how the function works—we need to learn how to use a function. More about this in Section 2.7 below about getting help.

Before we, at long last, run the round function, let's work out for ourselves the answer: 2.43:

```
round(x = 2.4326782647, digits = 2)
```

```
## [1] 2.43
```

Excellent. R got the right answer.

 Seriously, though—we just did something very important. We had a go at figuring out the answer *before* we asked R for it. This gives something for us to check R's answer against. This kind of checking is one of the very important ways to spot the insidious mistakes mentioned above: the ones that don't cause an error, but result in garbage results.

What do we need to do if we want to use the answer that a function gives us in a subsequent command? The answer is the same as above, when we wanted to keep the answer of some simple arithmetic for future use. We use the assignment arrow < - to assign the answer to a name. Let's do that, with a few other new wrinkles:

```
number_of_digits <- 2
my_number <- 2.4326782647
my_rounded_number <- round(x = my_number,
                           digits = number_of_digits)
```

Can you figure out what's going on here? The first two lines assign two numbers to a couple of names. Then, in the third line, those objects are given as the values of the two arguments. The answer is still 2.43

(of course), but we don't see the answer in the Console. It has been assigned the name my_rounded_number. Show yourself this—type in the Console my_rounded_number and press Enter.

- Functions do not alter their arguments. For example, the value of my_number was not altered to have two significant digits. Put another way, functions do not have 'side effects'; they just give their answer.
- We can give a function as the value of an argument of another function (e.g. round(x = sqrt(x = 7), digits = 2)). This results in functions nested inside each other, and generally is not so nice to read/understand, so we teach students to avoid it as much as possible.

2.5.1 FUNCTIONS, THE SEQUEL

Happy with functions? Good. Let's have a closer look. Check this out:

```
round(2.4326782647, 2)
```

```
## [1] 2.43
```

We didn't give the names of the arguments; we only gave the values. How can such magic work?! The answer is that round *expects* that we give the number it should round as the first argument, and the number of digits as the second. So if we don't tell it otherwise, by naming the arguments, it assumes the order/positioning of the arguments supplied. So, we don't have to name our arguments if we know their expected position.

And now check this out:

```
round(2.4326782647)
```

```
## [1] 2
```

More magic? How does `round` know to round to no significant digits? The answer is that some of the arguments of some functions have default values. The default value here is `digits = 0`. So we don't have to specify the value of a function if we know it has a default value and we're happy with that default value being used.

How do we know the expected position of the arguments, and how do we know about default values of functions? Again, we have to look at the instructions for how the function works—we need to learn how to use the function. More about this in Section 2.7 below, about getting help.

If you're not sure about the expected position of arguments, or about default values, look at the instructions for the function. You can then decide if you will name the arguments and if you can use default values. We use a mix of naming and position, and of giving values and using defaults—it's very situation-dependent. We will see lots of examples of both in the coming Workflow Demonstrations.

2.6 Add-on packages

An R package is a container for various things (functions, data) that make it possible to do special things with your own data; they are usually related by some topic. For example, there is a **stats** package that contains functions and other things for doing statistics. A specific type of thing contained in these packages is functions… for example, `mean` is a function to take the mean of some data. We need functions like trees need the sun and the world needs love. When you download R you get some pre-installed packages, like the **stats** one we just mentioned. There are, however, add-on packages that you can (and will need to) get and use functions from. So let's go through the process of finding packages, installing them, and using functions from them.

2.6.1 FINDING ADD-ON PACKAGES

Many add-on packages are available via the CRAN[7] website—yes, the same one from where we download R. CRAN stands for 'Comprehensive R Archive Network'. On the website you see about a dozen links in the left sidebar. In the *Software* section is *Packages*.[8] On the *Packages* page is a link to a table of available packages, sorted by name.[9] There are too many packages to search for one you might want, so what to do?

One answer is *Task Views*.[10] These are curated guides to packages and functions useful for certain topics/fields/disciplines. Have a browse through these. Some of them will make some sense; some will not. But now you know one place to go to learn about the various options for subject-specific tools in R. Some useful for life and environmental scientists include Environmentrics,[11] Experimental Design,[12] Graphics,[13] Multivariate,[14] Phylogenetics,[15] Spatial,[16] Survival,[17] and Time Series.[18]

We are now telling you about add-on packages because the approach we teach for using R requires some add-on packages. Specifically, add-on packages from what is called the tidyverse.[19] As we mentioned in the Preface, this is a collection of add-on packages and functions within them that, we believe, makes learning and using R easier, more efficient, and more pleasant.

---

[7] http://cran.r-project.org
[8] http://cran.r-project.org/web/packages/
[9] http://cran.r-project.org/web/packages/available_packages_by_name.html
[10] http://cran.r-project.org/web/views/
[11] http://cran.r-project.org/web/views/Environmetrics.html
[12] http://cran.r-project.org/web/views/ExperimentalDesign.html
[13] http://cran.r-project.org/web/views/Graphics.html
[14] http://cran.r-project.org/web/views/Multivariate.html
[15] http://cran.r-project.org/web/views/Phylogenetics.html
[16] http://cran.r-project.org/web/views/Spatial.html
[17] http://cran.r-project.org/web/views/Survival.html
[18] http://cran.r-project.org/web/views/TimeSeries.html
[19] https://www.tidyverse.org/

2.6.2 INSTALLING (DOWNLOADING) PACKAGES

To use a package and the functions in it, we need to do two things: (1) download it onto our computer (often termed 'installing' the package) and (2) load it into R's memory. Simple, eh?!

Most packages are available from CRAN, and to install them is quite simple (you must be connected to the internet!). For example, to install a package called **lubridate** type this:

```
install.packages("lubridate")
```

If everything worked, you will see some red text in the Console but no error message. You can also install more than one package at a time:

```
install.packages(c("dplyr", "ggplot2"))
```

A couple of things to note:

- Package names are case-sensitive. You will probably get an error now and again because you forgot this.
- Some packages need other packages (i.e. they have dependencies). When you install one, you may then find it installing others. Nothing to worry about—in fact this is quite useful.

 Install **dplyr** and **ggplot2**.

If you don't like typing, you're in luck. The RStudio Packages tab in pane 3 has an Install button at the top right. Clicking this opens a small window with three main fields: Install from, Packages, and Install to Library. Type the name of the package you would like into the Packages field; leave everything else with its default setting. As you type, RStudio will make suggestions, from which you can select the one you want, and than you click the Install button. RStudio does the rest (it runs the `install.packages` function).

Don't use `install.packages` *in scripts.* You only need to install a package once (i.e. you only need to download it once). But we might run our script hundreds of times. So don't waste time by having the package installed every time you run your script. You do, however, need to load the package into R's memory before you run your script, so it's a good idea to do that at the top of the script (see below for how).

We don't need to install the package (e.g. **dplyr**) every time we start a new R session. It is worth saying that again: *there is no need to install a package every time we start up R/RStudio.* Once we have a copy of the package on our hard drive, it will remain there for us to use. The only exception to this rule is that a major update to R (not RStudio) will sometimes require a complete reinstall of the packages. This is because the R installer will not copy installed packages to the major new version of R. These major updates are fairly infrequent, though, occurring perhaps every 1–2 years.

Installing a package does nothing more than place a copy of the relevant files on our hard drive. If we actually want to use the functions or the data that come with a package, we need to make them available in our current R session. Unlike package installation, this **load and attach** process, as it's known, has to be repeated every time we restart R. If we forget to load the package, we can't use it.

2.6.3 LOADING PACKAGES

OK, so we installed a package—it was downloaded onto our computer. Now we want to use it. Let's do this for the **dplyr** package:

```
library("dplyr")
```

You see that the `library` function is used to load the package. And you see some red text. AAAAAARGGGHH! Well, not quite. This is not an error or warning message. The first line tells us that R has loaded (attached) the package. This means that all the functions in the **dplyr** package are now available for us to use. The following lines tell us that some of the functions that R had in its memory are now 'masked' by new ones from the loaded package. This is generally nothing to worry about.

Don't use mouse clicks to load packages! You can load packages the clicky way (i.e. click RStudio buttons), but *do not do this*. Instead, always put the `library` statements in your script (but not `install.packages`!). They need to be run each time you open RStudio/run your script, so put them there. And put them all at the top of your script so all the required add-on packages are loaded before anything (or much) else is done. *Do not* have `library` statements dispersed throughout your code.

2.6.4 AN ANALOGY

Packages sometimes confuse new users, perhaps due to uncertainty about what the `install.packages` and `library` functions are doing. A useful analogy is with smartphone apps, in which an app is like an R package—it extends what a phone can do. To use an app we need to do two things: (1) install it and (2) open it. We only need to install it once (just like installing an R package). To use an installed app we tap its icon, and we have to do this every time we want to use the app. This is what `library` does.

2.6.5 UPDATING R, RSTUDIO, AND YOUR PACKAGES

Our advice is to only update things when you need to, and not every time a new version appears (i.e. don't be an 'early adopter'). If it ain't broke, don't fix it. If you want the latest and greatest versions, or if you get warning or error messages about versions being incompatible, you should think about installing updates. Otherwise, there aren't any hard and fast rules about when and when not to update any or all of R, RStudio, and your add-on packages.

Note that updating R does not automatically update RStudio (and updating RStudio does not automatically update R). If you want both updated, you need to update both. If you want one updated, you need to update that one.

To install updates of R and RStudio, go onto the websites previously described for getting these programs and get and install the latest version, exactly as you would if you were getting them for the first time. To update all your packages, use the Update feature in the RStudio packages tab.

Common add-on-package-related pitfalls include:

- Updating to the latest version of RStudio and thinking this also updates us to the latest version of R. Both must be independently updated. That is, update R. Then go and update RStudio.
- Thinking we need to install a package each time we want to use it. You only need to install once. You need to load the package at the start of each session, though (using the `library` function).
- Sometimes add-on packages fail to install, or give warnings (like that they were installed under some old version of R). Sometimes we get questions about whether we want to do something, with an answer yes/no. We usually flip a coin, and hit yes if it comes up tails and no if it comes up heads. If that doesn't work, we try the other answer! If that fails, we restart RStudio. If that fails, we update R, RStudio, and all our packages, then try again.

2.7 Getting help

There are many sources of information about using R: R itself, Google, RSeek, Twitter, Stack Overflow, manuals, books, friends, colleagues, instructors, online courses, books like this (and others mentioned on the *Insights* companion website).[20] Which one you need will depend on what you're seeking. If you know the function to use but are having trouble using it, look at the R help file (described in Section 2.7.1). If you want to browse what R can do in a particular subject area, perhaps try reading some relevant tutorial documents (known as vignettes; see Section 2.7.3). You can also ask questions in various online communities, e.g. the Stack Overflow website[21] (make sure you've looked first at the help files,[22] however, and thoroughly looked around on Stack Overflow and elsewhere for answers).

2.7.1 R HELP SYSTEM AND FILES

Every function has a manual telling us how to use it—these are what are frequently called the 'R help files'. As well as telling us how to use functions, these files also cover available datasets and perhaps even a tutorial (often referred to as a vignette). These help files are, unfortunately, anything but helpful to people starting with R, and especially folk that have not previously learned a programming language with the help of such documentation. They are concise reference documents—super-useful, standardized, but often a bit impenetrable, or at least a bit scary at first sight.

Sounds worrying. Let's take baby steps then. Open the help file for round by typing ?round in the Console—this will change the bottom right pane to the help tab and therein show the help file for round. Scan down to the *Usage* section. You will see a line round(x, digits = 0). This is where the help file tells us the expected order of the arguments and any default values of any arguments. Just below you will see the section *Arguments*, which contains a description of each possible argument.

[20] http://insightsfromdata.io
[21] https://stackoverflow.com/questions/tagged/r
[22] https://stackoverflow.com/help

The terms used are rather technical, but at least the description of the `digits` argument should make some sense. This is a starting point (more comes below).

If you keep looking down the help file for `round` you'll see the *See Also* section, with the entry `as.integer` as a hyperlink. Click on this and you'll go to the help file for the `as.integer` function. Help files and the Help tab are a bit like a mini web browser. Neat. There are forward and back arrow buttons for navigating forwards and backwards through the pages you've looked at. The home button takes us to a page with various useful sections, some of which are self-explanatory (e.g. Learning R Online) and some are not (e.g. CRAN Task Views). This home page is also likely to change, so we won't go through it in detail… but suggest you have a little explore, then come back here.

2.7.2 NAVIGATING HELP FILES

Let's take a closer look at the anatomy of a help file.

1. *Description*: a short overview of what the function is for. Sometimes a help file concerns a family of related functions, and then it gives an overview of the functions. Always read this section, and read it first.

2. *Usage*: how to use the function, or various functions if the help file concerns a family of related functions. Arguments without a default are listed on their own. Arguments with defaults are in name–value pairs, where the value is the default used if we don't give a value when we use the function. Always look at this section.

3. *Arguments*: the possible arguments, along with a brief description of what they do, what kind of data each argument needs, and the possible values (if there are a set of these). Always read this section.

4. *Details*: the details of how the function(s) behaves, often long and hard to understand. Worth scanning for stuff you understand, though—you will likely see bits you do, and then gain confidence and more and more understanding. Often, all we need to do is try.

That said, we can often ignore this section and not suffer. But when we want to understand the depth of a function, or suspect it might not be doing quite what we expect, we need to wade through this section.

5. *Value*: a description of the answer that the function gives us. Often possible to guess, but sometimes a bit unexpected. If we get back something we didn't expect, we probably need to learn a bit more about the function.

6. *References*: a list of key references. Dig into these to know the technical innards of a function. If the function implements a particular method/tool then a key reference about that tool should be here, and you should read it before using the tool.

7. *See Also*: as mentioned above, links to the help pages of related functions. If we know a function that nearly does what we want, we might find the one that does do what we want in this section. Also a nice way to browse functions, if you're into that kind of thing (a bit like browsing an encyclopedia).

8. *Examples*: no surprise here… examples of how to use the function. Seeing an example use of a function is often an excellent way to understand how it works/what it does. You can run these in R to see how they work.

2.7.3 VIGNETTES

A vignette is like a tutorial/walk-through document. Some of the better-developed parts of R have an associated vignette; others do not. Typing `browseVignettes()` in the Console and pressing Enter will show a list of available vignettes for the packages we have installed. These are great places to get acquainted with how to use R for different tasks. Each is a bit like a short book.

2.7.4 CHEAT SHEETS

Many add-on packages, and some topics in R, have cheat sheets, providing a compact and concise source of information about what one can do with the functions in a package, and how one can do it. They're not so good for beginners to learn from. But with just a little experience one can start to understand and use them. We recommend printing some out to look at and make notes on while working through *Insights*. In RStudio, go to the Help menu and down to Cheat Sheets to find some. Two to print out and work alongside during the Workflow Demonstration are *Data Transformation with dplyr*[23] and *Data Visualization with ggplot2.*[24]

2.7.5 OTHER SOURCES OF HELP

As well as the above, we quite often use a search engine, e.g. Google. If we were interested in how to round a number, we might type something like `R round number`. The first hit for us is the help file, then a web page of *R for Dummies*, and so on. If you want a bit more control than Google gives, use the search engine rseek.org.[25] It's pretty nice and only searches R-specific web pages. For the Twitter users among you, well, there is lots there about R… have a look at #rstats.

2.7.6 ASKING FOR HELP FROM OTHERS

This is a great way to get help. If you're a bit stuck, perhaps getting an error message, or just can't think of how to progress, and you've spent some time trying to help yourself, definitely ask someone (colleague, instructor, friend). When you do, it is usually very useful to show them your code and the error/problem you're experiencing. 'It doesn't work' is not very easy to diagnose and treat. 'Here is my code, here is some of the data, here is what I want it to do, and here's the problem' is much easier to help with. Make sure your code runs and recreates the problem. Also, try to keep it simple for the person to see and understand. In sum, attempt to make what is known as a 'reproducible example' (the **reprex** package can help make these, and here are nice instructions for making a reproducible example[26]).

---

[23] https://github.com/rstudio/cheatsheets/raw/master/data-transformation.pdf
[24] https://github.com/rstudio/cheatsheets/raw/master/data-visualization-2.1.pdf
[25] http://rseek.org
[26] https://reprex.tidyverse.org/articles/reprex-dos-and-donts.html

It may be daunting to send your code to someone for help. You might think it is not such good code, or even really bad code. This is a natural and common feeling. Perhaps share first with someone with whom you can trust your feelings. Then maybe take a bit of a risk of sharing with someone else. We remember doing this… it was scary, and then was OK, and later it was even funny to see the mistakes we made, and still make.

2.8 Common pitfalls

- Updating to the latest version of RStudio and thinking this also updates us to the latest version of R. Both must be independently updated. That is, update R. Then go and update RStudio.
- Thinking we need to install a package each time we want to use it. We only need to install once. We need to load the package at the start of each session, though (using the `library` function).
- RStudio very frequently 'autosaves' your script files. So if RStudio stops responding (a rare but not unknown event), first try to save your script in RStudio… this may work. If you're not sure this worked, and can still highlight and copy your script, then do so, and paste it in to a text editor application. Then kill RStudio and restart it. When it restarts, you will very likely see the script you were just working on, as you left it. If not, paste in the code you copied to the text editor and carry on.
- The dreaded plus sign in the Console is so annoying when it happens that we thought we'd remind you about it (see Section 2.4.3).

2.9 Summing up and looking forward

You probably feel like you experienced a lot in this chapter, and maybe didn't quite follow it all. That's fine and to be expected, in large part because you were experiencing quite a lot of quite new tools and concepts. Learning multiple new things all at once is a challenge, but we think the integrated understanding that comes from learning this way is worth the challenge and even a bit of temporary frustration.

It could be worth now getting together with someone experienced with R and RStudio and asking them about things you're uncertain about. Be aware, though, that there are many different types of R user: if the one you find seems to do things very differently from what we've shown you (e.g. does not use RStudio or the tidyverse), it could be hard going. Perhaps find someone else!

You hopefully also feel that you accomplished a lot. You got acquainted with your toolbox, tried out some of the simple tools, and made a script that you can use to tell your tools what to do and to save those instructions. You learned how tools can go wrong, and what to do when they do. You learned about the structure of and how to work with a type of tool called a function. You learned how to get more tools (add-on packages) and how to get help about using your tools. And about a few common pitfalls to avoid. This is all a fantastic foundation upon which to experience a demonstration of getting insights from real data, which is the subject of the next chapter.

Workflow Demonstration part 1
Preparation

We hope you are now acquainted with some core skills associated with using R and RStudio. These will make life easy and allow you to focus on the task of data manipulation and visualization, and insight is really important here.

This and the next chapter involve us working through the journey of getting insights from data. We will step into the shoes and mind of a researcher, and try to experience what they would during the latter stages of a research project, where data have been collected against some preconceived ideas/hypotheses. While you didn't collect them, you will download the data and will pretend, alongside us, that they are our own hard-won data, with sweat and tears shed while collecting them, the drudgery experienced of typing them in, and the anticipation mounting of finding out what they contain—insights, however small or large.

We have chosen a research subject that should have some interest to us all: food. More specifically, what animals eat. Even more specifically, what bats eat. Isn't that exciting? Bats are cool. We chose this study as it is not too simple and not too complex. It is simple because there are relatively few variables to consider and little correlation among explanatory variables—

Insights from data with R: An Introduction for the Life and Environmental Sciences. Owen L. Petchey, Andrew P. Beckerman, Natalie Cooper and Dylan Z. Childs, Oxford University Press (2021). © Owen L. Petchey, Andrew P. Beckerman, Natalie Cooper and Dylan Z. Childs. DOI: 10.1093/oso/9780198849810.003.0003

both of these make drawing conclusions and getting insights simpler. There are also a reasonably large number of data points (individual bats), such that each individual data point has relatively little influence on the overall patterns.

However, there are some specific and rather common problems in these data. Problems that one often needs to work through before even starting to try the tools and techniques that offer insight. There are problems with the names of variables. There are problems with the presentation of the values recorded for some of the variables. There are issues with a variable that represents dates. Further complexity comes from having to calculate one of the variables we are trying to explain—it's not encoded yet in the dataset. Not to mention that there are actually three separate response variables (the things we are most interested in understanding) to examine. Finally, the data were collected via observation; there were no manipulations/interventions in the study, which further adds an issue of correlation among some of the variables.

 Print the workflow! To ensure you know exactly where in the workflow we are, and if you have not already done so, please print out a copy of our workflow, either the diagram or the checklist, both of which are available at http://insightsfromdata.io. As you work through this chapter, keep an eye on where in the workflow you are, and make notes on that section.

To deal with all of this, and start on the insights journey, you will:

- Get to grips with a clearly specified set of research questions.
- Learn how the study was performed and why it was done that way.
- Get the dataset from an online repository.
- Learn how to prepare your computer, R, and RStudio for the project and tasks.
- Learn how to read the dataset into R, and then how to check the import.

- Learn how to clean, tidy, and manipulate the data to allow tables and graphs to be produced.

As we go through the demonstration, we will learn quite a bit of R. We will also learn some concepts central to insights and statistics (e.g. independence of observations). The material in the demonstration does, however, leave many things to learn until later. So do not worry if you don't understand all of the new R and new concepts. They are covered again, in more detail, in subsequent chapters. The main point, at this point, is to experience and familiarize yourself with the entire process of getting insights from data.

3.1 What is the question?

Imagine that you have, for as long as you can recall, been interested in bats. And that you recently came across a study of the differences in the diets of male and female bats. Perhaps male and female bats eat different things? We might expect this if, for example, male and female bats are different sizes. And perhaps adult and juvenile bats eat different things?

With new molecular methods, we can identify the food eaten by bats by analysing the DNA in their poop. So you apply for and are awarded funding to collect bat poop and analyse the prey in their diet, by molecular analyses of the poop. You propose to look for sex and age differences in the diversity, size, and migratory nature of the prey. Put another way, you aim to see if each of these three *response variables* differs between male and female bats, and if these differences depend on the age of the bat.

We think it is useful to see now what we are aiming for: a classy visualization that captures our question and allows an attempt at insight. Have a look at Figure 3.1; it's one of the graphs you'll make—a visualization of the prey size data for male and female bats that are juvenile or adults. Can you see a pattern? You're on your way!

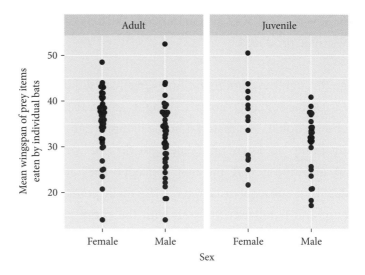

Figure 3.1 Difference in prey size (measured as wingspan of the prey item) among male and female bats that are adult or juvenile. This graph is just to show you what we are aiming for, and to clarify in your mind the question of this Workflow Demonstration. We will improve this graph, for example by reducing the extent to which individual data points are plotted on top of each other.

3.1.1 THE THREE RESPONSE VARIABLES

It's important to recognize that we have one overarching question—do bat diets differ by sex and age? Embedded in this overarching question are three sub-questions, specific representations of that overarcher that arise from focusing on three specific variables—response variables—that were measured: prey diversity (number of prey), prey size, and the migratory nature of the prey. Let's review each of these response variables in turn.

Number of prey species consumed by a bat. This varies between 0 and the total number of species that a bat might eat (though clearly no single bat is going to eat all of them). Values are *integer* (whole numbers only) and *cannot be negative*. And when we make a single observation of the diet of an individual bat, the measured number of prey species can only be integer (an individual bat cannot eat half a prey species: even if it eats only half an individual, it is still counted as having that whole prey species in its diet). Hence these are **count** data… they can only be integer and non-negative.

(Note that this is the number of species, and not the number of individuals, i.e. while a bat might eat two individuals from the same species, the number of prey species eaten would still be one.)

Information. We have just explained the type of variable the number of prey species is. We also do this for the other two variables just below. We do not go into great detail during this workflow about types of variables. You should now choose whether you will continue with the workflow, or have a look at a section of a later chapter that details types of variables (Section 8.1). You will have to make this choice to look ahead or not multiple times while going through this demonstration. We suggest you continue through the demonstration, making notes about what you are unsure of, and later look in more detail.

Average size of prey consumed by a bat. This is a *continuous numeric variable*. In this particular study it is the average wingspan, in millimetres, of the prey species eaten by a particular bat.

Proportion of prey that are migratory consumed by a bat. For each prey item, this is a *binary variable* that can be either yes, the species is migratory, or no, it is not. That is, each of the prey species eaten by a bat can be migratory or not. The maximum proportion of migratory prey in the diet of a particular bat is 1, and the minimum 0.

3.1.2 THE HYPOTHESES

'[I]t might be expected that owing to their high energetic requirements during breeding, females should feed more than males on large prey and on prey with high energetic value such as migratory moths'. This is a quote from the scientists that conducted the research. It is quite a nice hypothesis, as it has a mechanistic foundation (female bats have higher energetic requirements) and is directional (females feed more than males). A weaker hypothesis could be, for example, non-directional (females and males have different diets). The scientists did not present hypotheses about the number of prey species or their migratory nature, so their analyses of these were more exploratory.

3.2 Design of the study

While we've been working under the assumption that we have the data, it's important to recognize and understand how the data were collected. It's very likely that these data were collected after the hypothesis above was formulated. Some practicalities must have been sorted out, including:

- Which bat species to study?
- Should data be collected from feeding trials on wild or captive bats or observed from the natural diet of wild bats?
- Where and when should the work be done?

These issues don't directly impact on our understanding of the data. However, other questions about design do, and we have discussed these issues already in Section 1.3:

- How many bats and how many males and females were measured?
- What will be considered the independent observations?
- What are the likely sources of uncertainty and variability? Are there many observers doing the work? Is the machine for poop analysis accurate?
- Have any manipulations been done?
- What specific variables will we record to address the hypotheses, and are they at all (likely to be) correlated?

The number of bats and the question of what will be the 'independent observations' are vital. A clearly specified question will often help with this. We are asking if the diets of adult or juvenile, male or female bats differ. Assuming we don't measure the same bat as a juvenile *and* adult, our unit of replication must be the diet of an individual bat. Put another way, our graphs showing the data should have the same number of data points as the number of bats we caught.

3.3 Preparing your data

Imagine now that we have completed the practical part of the study, and have our data. The data are distributed in various places, perhaps some in a spreadsheet, some on paper, and some in files provided by machines. We need to organize this into a single spreadsheet. This can be quite a lot of work, and is very bespoke work. It is possible to use R to do some of it, but it also might also be as simple as typing data in by hand.

Very importantly, we need to know how we'd like our data to look once they are organized. We strongly recommend that you organize your data in what is often referred to as *tidy* format. This is because the tidyverse-based approach for getting insights from data in R is most powerful when data are tidy.[1] It's also pretty much the easiest format for computers to read.

Here 'tidy' implies a specific structure that makes it easy to manipulate and visualize the data. In a tidy dataset, each *variable* is in only one column and each row contains only one *observation*. Cells of the table/spreadsheet contain the *values*. It's nice that you are familiar with the ideas of *observations*, *values*, and *variables* from previous chapters!

When you type in data, use your favourite spreadsheet program (e.g. Excel, Numbers, Libre/OpenOffice). If we were typing in the bat diet data, we would make each row contain information about one observation, of one prey species in a poop, from one bat (Figure 3.2). You are probably thinking 'that's a huge amount of repetition!' and you are right. But this also allows us to use the computer to make summaries at any level we want. Keep in mind that if the different prey were spread across multiple columns, we would have 'untidy' data, and then would need to do some tidying. We come back to this when we explore the bat diet data.

One you've typed in your data, and double-checked them against the original data, save them as a 'comma-separated values' (CSV) file type.

---

[1] http://vita.had.co.nz/papers/tidy-data.pdf

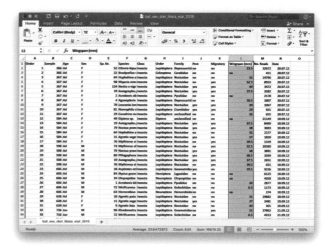

Figure 3.2 How the bat diet would appear when we had finished typing it into a spreadsheet.

```
⬤ ⬤ ⬤                    ⬆ bat_sex_diet_Mata_etal_2016.csv
Order,Sample,Age,Sex,Sp. Nr.,Species,Class,Order,Family,Pest,Migratory,Wingspan (mm),No. Reads,Date
1,366,Ad,F,52,Ethmia bipunctella,Insecta,Lepidoptera,Depressariidae,no,no,23.5,32672,20.07.12
2,366,Ad,F,22,Bradycellus verbasci,Insecta,Coleoptera,Carabidae,no,no,na,411,20.07.12
3,367,Ad,F,64,Hoplodrina ambigua/ superstes,Insecta,Lepidoptera,Noctuidae,no,no,31,14796,20.07.12
4,367,Ad,F,98,Rhyacia simulans,Insecta,Lepidoptera,Noctuidae,no,yes,52.5,8953,20.07.12
5,367,Ad,F,114,Xestia c-nigrum,Insecta,Lepidoptera,Noctuidae,yes,yes,40,3472,20.07.12
6,367,Ad,F,19,Autographa gamma/ pulchrina,Insecta,Lepidoptera,Noctuidae,yes,yes,37.5,3282,20.07.12
7,367,Ad,F,2,Acrobasis obliqua,Insecta,Lepidoptera,Pyralidae,no,no,na,2178,20.07.12
8,367,Ad,F,4,Agonopterix scopariella,Insecta,Lepidoptera,Depressariidae,no,no,20.5,1807,20.07.12
9,367,Ad,F,70,Leucania loreyi,Insecta,Lepidoptera,Noctuidae,yes,yes,39,1067,20.07.12
10,367,Ad,F,81,Nomophila noctuella,Insecta,Lepidoptera,Crambidae,yes,yes,29,925,20.07.12
11,367,Ad,F,29,Caradrina morpheus/ Hypena sp./ Eupithecia sp.,Insecta,Lepidoptera,unclassified
Lepidoptera,no,no,35,655,20.07.12
12,584,Ad,F,49,Diptera sp.,Insecta,Diptera,unclassified Diptera,no,no,na,21149,10.09.12
13,584,Ad,F,19,Autographa gamma/ pulchrina,Insecta,Lepidoptera,Noctuidae,yes,yes,37.5,8499,10.09.12
```

Figure 3.3 How the CSV file of bat diet data looks when we open it in a plain text editor.

These files just contain plain text (no coloured cells, no formulae, no bold or italic text). Figure 3.3 shows how the bat data CSV file looks when you open it in a text editor. Each row is a row in the data. Each value in a row (corresponding to a different column) is separated by a comma. Variable names are in the first row; this is sometimes termed the header row.

It is much more convenient to type data into a spreadsheet than as plain text. We need to be careful, however: Excel has a nasty habit of formatting (changing the appearance of) data values. While it is possible to import data directly from an Excel worksheet into R, we recommend against this. Instead, we advise you first export the relevant worksheet to a CSV file, and then import this CSV file into R. This involves opening the relevant Excel worksheet, going to `Save As...`, choosing the `Comma Separated Values (.CSV)` file format, and pressing Save. Now we are free from Excel. *Freedom!*

Is it really worth making a CSV version of the Excel file? 'Probably' is the answer. We can be certain that we'll always be able to read a CSV-formatted file (the format is not proprietary, for example). We can't be so sure about an Excel file. Probably you can remember a time when you couldn't open an old file because you no longer had the necessary software (or you would remember such a time if you were as old as us). You can avoid this possibility by using the CSV format. So yes, we think it is worth it. Especially if you're a little paranoid, and we hope you are in your data analyses.

Excel will often cause problems. Many problems with importing data into R have their origin in Excel! For example, imagine we have data in Excel with three columns. If we accidentally add a value to a cell in the fourth column and then remove it, Excel may continue to consider the fourth column to contain data. We then end up with four columns in the CSV file and in R, where the fourth column is blank. The same thing can happen with empty rows. This isn't too difficult to remedy once the data are in R, but it is annoying and unnecessary work created by Excel.

Excel will often cause problems (yes, again). Another problem Excel sometimes creates is changing the separator from a comma to something else. That is, if you open a comma-separated text file in Excel and allow Excel to then save it, Excel may change the separator to something else (e.g. a semicolon) even though you did not ask it to.

Then you will need to change the separator when you read it into R, but not before you get frustrated because it wouldn't import the first time!

Sharing data. Now, in the workflow, is a good time to begin preparing the data for sharing. This would include, for example, writing metadata to accompany the data. By the way, the header row of a data file contains metadata already... the variable names. More information about preparing data for sharing is in Section 10.2.

3.3.1 ACQUIRE THE DATASET

We briefly now stop pretending we are the researcher, as we need to get the data and know what they are. The data come from a study entitled 'Female dietary bias towards large migratory moths in the European free-tailed bat (*Tadarida teniotis*)', published by Mata and colleagues in 2016 in the journal *Biology Letters*.[2] The authors were good enough to also publish the data.

We get the dataset `dataset_Mata.et.al.2016.xlsx` from the folder 'Diet analysis dataset' from the Dryad repository[3] associated with the publication about the data. Please ensure you get the file from *version 2* of the Dryad data publication. The dataset is stored on Dryad in `.xlsx` format.

As noted above, we prefer to read CSV files into R, so before starting working in RStudio, we need to convert this file into a CSV format. Once you've downloaded the file, open it in Excel or your favourite spreadsheet program. Now, use the File >Save As menu to change the file type to a CSV file. Make sure to call it `bat_sex_diet_Mata_etal_2016.CSV`. Make sure you know where this CSV file is saved; at this point, it's probably in your Downloads folder or desktop.

---

[2] http://rsbl.royalsocietypublishing.org/content/12/3/20150988
[3] https://doi.org/10.5061/dryad.m8t72.2

Feel free now to delete the `.xlsx` file, if you wish—we won't use it again. See earlier in this section for our reasons for using CSV files and avoiding `.xlsx` ones as much as is humanly possible.

Let's now open the CSV-formatted dataset in a spreadsheet program such as Excel, so we can take a look. We see a dataset with 633 rows and 14 variables. If we had entered the data ourselves, we would understand their structure. As we didn't, we need to get this understanding embedded before we start working with these data.

Here is our interpretation of the 14 variables in the dataset:

- *Order*: appears to be row number; goes from 1 to 633.
- *Sample*: this is a number indicating the individual bat that the row of data concerns.
- *Age*: age of the bat, as juvenile or adult.
- *Sex*: sex of the bat.
- *Sp. Nr.*: A numeric code for the species of prey a row concerns.
- *Class*, *Order*, and *Family*: taxonomy of the prey.
- *Pest*: is the prey a pest species?
- *Migratory*: is the prey migratory?
- *Wingspan (mm)*: the wingspan of the prey species.
- *No. Reads*: number of reads for this prey in the molecular analysis.
- *Date*: the date the poop was collected on.

So, *Sample* is the identifier for a particular bat individual (it might have been more intuitive to call this variable `Bat_ID` or something similar, and we will rename it to this later). Each individual bat can appear on multiple rows, with each row being for a particular species of prey found in its poop. Hence, each row corresponds to a particular observation of a bat eating a prey species (i.e. a predator–prey interaction). As you can now see, a collection of rows for any individual bat gives the diet of that bat. Alternatively, the collection of rows for a particular prey would be all the bat individuals that ate that prey species.

By the way, reflecting on the list of things we noted might be wrong with the data, you may have now spotted the potentially problematic issue that there are two columns with the same name (`Order`) and some of the column names have spaces. These are problems you will encounter in the future. They are fixable in R very easily, and we will do this such that our script contains a record of all manipulations we perform.

Now, with a look again back to the variable names, let's have a think about how these data could be used to answer our three research questions. Only the `Wingspan (mm)` variable, representing prey size, seems a direct representation of one of our intended response variables (diversity, size, and migratory nature). This means we will have to do some work to get the other two. Can you envision any of the calculations we might need to do in order to get these variables?

OK. We've now got a reasonable idea about the data we're going to use. Let's organize our computer now to use these data with R.

3.4 Preparing your computer

Back to pretending we are the researcher... We have our question, we've performed our study, we've typed in the data and made sure they're a CSV file—we definitely know where the file is on our computer, and we've backed it up. Haven't we?

But we are not quite ready to start up R. The next step of our workflow is to set up our project on our computer. What constitutes a project? Well, anything asking a reasonably coherent set of questions using one, or sometimes more than one, dataset. For example, you might have a coursework, practical/lab, or honours project where you're analysing data collected in a lab or in the field. Perhaps you've had a summer bursary to do some research. Perhaps you are doing a masters or PhD, with multiple chapters, and then each chapter is a project. Finally, you might be working on a brilliant manuscript (they all are!)—our manuscripts are treated as projects.

A complementary way to define a project is by 'product', or set of products. A product is something of value for others, e.g. a report for colleagues, an article for the public, or a presentation at a meeting. This view of projects encourages us to work with the end in mind, and may help us prioritize activities (i.e. which project/product does this activity contribute to and how does it contribute?).

Projects will likely contain several files, including data and R scripts. You may also want to store information related to the report/thesis chapter/-coursework that you're writing. A good place to start is by creating one easy-to-use and easy-to-find place (folder) to keep all this information together for each of your projects.

We all have different ways we like to do this, but the one we recommend is to have a folder called Projects on your desktop or somewhere like your Documents folder that is easy to find on your computer; yes, you can use Dropbox, Google Drive, SharePoint, etc.

3.4.1 MAKING THE PROJECT FOLDER FOR THE BAT DATA

Now, within that folder make a subfolder for this specific project. Create a project folder called *bat_sex_diets* in this master folder in your Documents or desktop, or some other location on your computer. Make sure there is a data folder and a scripts folder inside this.

Next, within this (and any future) *bat_sex diets* folder it's good to further organize yourself by having subfolders called 'data', 'scripts', 'figures', etc. This helps to tidy everything up, and means it's harder to lose important components of the project and analysis than if you've got things floating all over your computer! Let's do this for our bat diet project.

Now, do you remember where `bat_sex_diet_Mata_etal_2016.CSV` is on your computer? We know you do.... go and drag/move the dataset into the data folder inside the *bat_sex_diets* project folder. (If a picture might be useful, skip forward to Figure 3.5—it shows our folders after we set up this project and a few others.)

Phew! Quite a bit of work, but it's worth it.

3.4.2 PROJECTS IN RSTUDIO

Once the dataset is in your data folder in the *bat_sex_diets* folder, we can use RStudio to declare this folder as containing a *project*.

To make a project for the bat diet project:

1. Look around RStudio, find the project icon (i.e. Figure 3.4), and click on it (any will do). Depending which one you clicked, you may then need to click New Project...
2. You should now have a dialogue box with three options: *New Directory*, *Existing Directory*, and *Version Control*. (Note that *Directory* is the word used to mean a *folder* on your computer.) Because we already have a folder for the bat diet project, choose the option *Existing Directory*.
3. The dialogue box moves on and asks what type of project we want. Choose *New Project*.
4. Then we can then use the *Browse* button to locate our folder bat_sex_diets. This is the folder on your computer in which the project will reside, i.e. the one we made that contains the data.
5. Finally, click the *Create Project* button.

A few things then happen:

- RStudio may ask if you want to save any unsaved files.
- RStudio looks like it restarted.
- A new file is put in the *bat_sex_diets* folder: the Rproj file, with the same file name as the folder/Rproject.
- A bunch of other useful stuff also happens, but we won't look at that right now.

Figure 3.4 The icon to click to create (right) or switch to (left) a project in RStudio.

Creating the .Rproj file, in addition to the details above, offers a really big advantage: it declares to R that you are working in a very specific place on your computer (in the *bat_sex_diets* project) and your data are in this place for this project. This makes it easy for R to import the data (see next chapter) that we need.

Repeat the process for the three other Workflow Demonstration studies (find them on the *Insights* companion website),[4] perhaps naming the project folders polity_food_diversity, diet_div_pop_stab, and fish_diet_restriction. Then, make sure there is a data and scripts folder inside each, and go ahead and make the project file too. You should then have four folders, one for each study, each containing its own .Rproj file (e.g. as shown in Figure 3.5).

Excellent! This is major progress. After we have done all this, when we click on the Project menu in the top right of the RStudio window we should see our Project bat_sex_diets listed (as shown in Figure 3.6).

That was really important, so it's worth saying again. Check out the project icon in the top right of the RStudio window. Next to the icon is the name of the project you're currently working in, and next to that

Figure 3.5 This is how our folders looked when we created our four project folders, and their contents (only the contents of the first Workflow Demonstration folder are shown).

---

[4] http://insightsfromdata.io

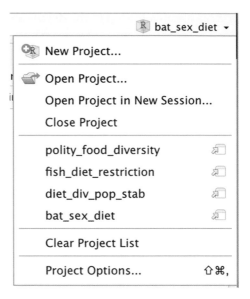

Figure 3.6 The projects menu in the top right of the RStudio window is a really convenient place to see and switch among your projects (as well as do other project-related things).

is a little black triangle, which means something will happen if you click it. Click it. Various options appear (what we see when we click it is shown in Figure 3.6), including to make a new project, and also the other projects you were previously working in. The implication of this is that you can now very easily switch between your projects (you don't even need to know where their folder is, but it's still important to know). You even get a warning if you haven't saved work in your current project before asking to switch to another.

You will find that organizing your work into projects like this is immensely valuable. If you like, you can get the empty folders and project files ready made from the *Insights* companion website.[5]

---

[5] http://insightsfromdata.io

3.4.3 CREATE A NEW R SCRIPT AND LOAD PACKAGES

A few final bits of preparation… let's now create a new R script file in which we will write, annotate, and store instructions for manipulating the data and developing tables and graphs for insights. Use the icons, or the File > New Script menu item.

Once this opens, add a few lines of comments that describe what will be in the script, for example

```
# An analysis of the diets of male and female bats.
# Data from "Female dietary bias towards large migratory moths
# in the European free-tailed bat (Tadarida teniotis)"
# By Mata and colleagues in 2016 in the journal Biology Letters
# http://rsbl.royalsocietypublishing.org/content/12/3/20150988
```

Now, add the following code, which will load the add-on packages we will use into R's memory:

```
# Load the libraries we use
library(dplyr)
library(ggplot2)
library(readr)
library(stringr)
library(lubridate)
library(tidyr)
library(ggbeeswarm)
```

Now, save this file into the scripts folder inside the *bat_sex_diets* project folder: call it *bat_sex_diets.R*. If, when you are saving, you don't see the .R at the end of the file name, add it.

Note that you now also need to *Install* the **dplyr, ggplot2, readr, stringr, lubridate, tidyr,** and **ggbeeswarm** packages; this must be done before you try to use the code in your script, otherwise the library(***) calls (where *** is one of the package names) will not work.

Look at Section 2.6 if you need help with installation.

Great! You now have a project folder called *bat_sex_diets*. In there, you have a folder called 'data', with the file `bat_sex_diet_Mata_etal_2016.CSV` in it. We also have in the project folder a *bat_sex_diets.Rproj* file that has declared to R that this is a project folder—we are in there, and R is looking in there too with us. Love at first R.

We also have a scripts folder, where we have the *bat_sex_diets.R* script with some annotation and a section on the libraries we need. We are ready to import the data into R's memory now. Woot!

3.5 Get the data into R

We have our question, and we have performed our study, typed in the data (well, downloaded them), organized a folder for this project, and prepared a project. Now it's time to *import* the data into R.

Open your project. If you haven't already got the `bat_sex_diet` project open, double-click on your project file (we called it *bat_sex_diet.Rproj*) to open the project in RStudio.

Now let's get the data into R. So, just checking: you've got your project, in it made a `data` folder, and in it put the CSV file that you made from the file you downloaded. Make sure you opened RStudio by clicking on your project file for this Workflow Demonstration (or opened the project from a menu). This will ensure RStudio is looking at the correct folder, and that the *relative path* used before the file name will be correct.

We just used the term *relative path*. What is this? The word *path* means the path, or directions, to where the data file is, i.e. directions through the folders on your computer. And *relative* means directions that start from where we are currently. An absolute path would be like giving directions always from one place, e.g. the meeting point at the main station in Zürich. Relative paths, in combination with RStudio projects, make the directions to our data files independent of where we (or a team member) put the project folder. We can, for example, move the project folder from a 'first draft' to a 'submitted' project folder when we submit it (should we do such things), without having to change any paths in our R scripts.

Now, add the following code to the script, below the libraries section:

```
# read in the data from the data folder in my project
bats <- read_csv("data/bat_sex_diet_Mata_etal_2015.csv")
```

```
## Error: 'data/bat_sex_diet_Mata_etal_ 2015.csv' does not
exist in current working directory ('/Users/owen/GitHub/...
```

This code is designed to load the dataset using the `read_csv()` function from the **readr** add-on package.

In order for all of this to work, you need to put the cursor on the first `library()` line and run the code from there down. You can do this using the Ctrl-Enter (Windows) or the Cmd-Enter (Mac) keystroke combination. This sends the instructions from the script into R's brain. Do this now.

Whoops! Houston, we have a problem. We get `Error: blah blah blah does not exist in current working directory (blah blah blah)`. This usually happens if we wrote the wrong file name and/or path. The mistake above is that we have 2015 in the file name when it should be 2016. Let's edit and try again...

```
# read in the data from the data folder in my project
bats <- read_csv("data/bat_sex_diet_Mata_etal_2016.csv")
```

```
## Warning: Duplicated column names deduplicated: 'Order'
## => 'Order_1' [8]
```

```
## Parsed with column specification:
## cols(
##    Order = col_double(),
##    Sample = col_double(),
##    Age = col_character(),
##    Sex = col_character(),
##    `Sp. Nr.` = col_double(),
##    Species = col_character(),
##    Class = col_character(),
##    Order_1 = col_character(),
##    Family = col_character(),
##    Pest = col_character(),
##    Migratory = col_character(),
##    `Wingspan (mm)` = col_character(),
##    `No. Reads` = col_double(),
##    Date = col_character()
## )
```

Great! We didn't get an error. The `read_csv()` function has done two things. R has read in the data, and the `bats <-` assigns the data to the name `bats` (you could give the data a name other than `bats` if you wish). We cover in detail what this object is, and how data are stored in it, in Section 3.6.1. If you look in the upper right of RStudio, at the Global Environment box, you should see the word `bats`.

The `read_csv()` function also gives a summary of what it found and imported. We get a warning message: `Warning: Duplicated column names deduplicated: 'Order' => 'Order_1' [8]`. We have already noticed that the dataset contains two variables named `Order`. We also see that `read_csv()` has *deduplicated* these by adding `_1` to the name of the second variable. Thanks, `read_csv()`!

The two lines `Parsed with column specification:` and `cols(` are always the same, and just tell us what is coming—a list of the columns/variables the function found in the data file, and the type of variable it has made them be in R. So, the first variable it found is `Order`, and `col_double()` means that R sees that the variable contains numbers, some of which are fractions. The word `Order` comes from the first row of the data file `bat_sex_diet_Mata_etal_2016.csv`, which contains the column/variable names (sometimes known as the *header row* or *header*). The third variable is `Age` and is of type `col_character()`, meaning that it contains some letters/words. The fifth variable name contains a space, so R has surrounded its name in single quotes, i.e. `'Sp. Nr.'` (the same has happened with two other variables). (If you have already noticed that some of the variables seem to be of incorrect type, great. We will explain and fix this below.)

> Please use the `read_csv` function (from the **readr** add-on package). Please do not use `read.csv`, as it has slightly but importantly different behaviour that we do not want.

> *Importing data from other sources.* R is able to read data from many different data storage types, formats, and locations. It can get data from various databases, from Excel and other proprietary formats, from websites, and from good-ole text files. We deal mostly with text (e.g. CSV) files because they're simple and, in the majority of cases, quite sufficient.
>
>   In the food-diversity-polity Workflow Demonstration we will use a different file format to store an intermediary dataset, and also show how one can, if one really wants to, read directly from an Excel file. Oh, and by the way, there are straightforward methods for reading data directly from a Google Sheet (the Live Data Analysis Demonstration detailed on the *Insights* companion website[6] includes reading data from a Google Sheet that has been populated by people entering data into a Google Form). That's just a few of the many ways of getting data into R. We only need one for now, and mostly only use one ourselves: reading in a CSV file using `read_csv`.

---

[6] http://insightsfromdata.io

3.5.1 VIEW AND REFINE THE IMPORT

Finally we have our beloved data in R! This is great, but we now need to
check they have been appropriately imported. But what does 'appropriately
imported' mean? It means 'did we import what we meant to?' For example,
does it have the correct numbers of rows and variables, and are the variables
of the expected type (e.g. character, numeric)? We can look at these features
of the imported data in many ways. One useful method is to use the
glimpse function on the bats object:

```
# check the structure of the data
glimpse(bats)
```

```
## Rows: 633
## Columns: 14
## $ Order          <dbl> 1, 2, 3, 4, 5, 6, 7, 8, 9,...
## $ Sample         <dbl> 366, 366, 367, 367, 367, 3...
## $ Age            <chr> "Ad", "Ad", "Ad", "Ad", "A...
## $ Sex            <chr> "F", "F", "F", "F", "F", "...
## $ 'Sp. Nr.'      <dbl> 52, 22, 64, 98, 114, 19, 2...
## $ Species        <chr> "Ethmia bipunctella", "Bra...
## $ Class          <chr> "Insecta", "Insecta", "Ins...
## $ Order_1        <chr> "Lepidoptera", "Coleoptera...
## $ Family         <chr> "Depressariidae", "Carabid...
## $ Pest           <chr> "no", "no", "no", "no", "y...
## $ Migratory      <chr> "no", "no", "no", "yes", "...
## $ 'Wingspan (mm)' <chr> "23.5", "na", "31", "52.5"...
## $ 'No. Reads'    <dbl> 32672, 411, 14796, 8953, 3...
## $ Date           <chr> "20.07.12", "20.07.12", "2...
```

When we run glimpse(bats) we get several lines of output. The
first two give us the number of observations (rows) (633) and number of
variables (14). This is what we expect if we look at the data file. All looks
good so far, then.

In the next lines we see variable names and types, and some entries.
But all is not well. Woe be us! Some of the variables that should be

numeric are character. For example, we see `Wingspan (mm)` is followed by `<chr>`, meaning it is a character (string) type of variable; it is reasonable to assume that this variable should be numeric (more about strings in Section 6.2). Furthermore, the final variable `Date` is imported as `col_character()`, meaning R thinks it contains words, rather than recognizing the data are dates.

Another method to view the data is to type `View(bats)` in the Console and hit Enter. This will open a spreadsheet-like view of the data. You can also click on the blue circle next to the `bats` word in the Global Environment in the upper right—this shows a version of `glimpse`—or click the word `bats`, which will also spawn the spreadsheet view. Another is to type `bats` in the Console, hit Enter, and get another view of the data. In this book we switch between methods a bit, hopefully using the most appropriate one for a particular situation.

View only displays the data. The `View` function shows us the dataset in what looks like a spreadsheet. We can even reorder the rows by a column, by clicking on a column name. We can't, however change the dataset—this is very good, as we only want changes to be made with code, so we have a record of changes made.

Which method should you use for viewing features of the dataset in R? We tend to use each at different times. There is not a wrong or right method. Try these out and see which works best for you. But never forget the safety check of making sure the data you wanted were imported into R's brain.

Factors. You may or may not have heard about *factors*. These are special types of variable that contain words (strings). One reason that we use `read_csv` is that it does not automatically create factor-type variables, which its close relative `read.csv`, used to do by default. This is no longer a problem in the newest version of R.

3.6 Getting going with data management

We've imported the data, safety checked that it is what we expected, but already found some inconsistencies. We anticipated having to deal with some variable names, but now we can see even more. Here we start using some of the core functions that the tidyverse set of packages offers us.

If what R thinks is a character variable should in fact be numeric, it probably contains some words as well as numbers. So let us have a look at the `Wingspan (mm)` variable. We use the `select` function to look only at this variable, and then the `head` function to specify the number of rows to look at. Note that because the variable name contains a space, we must surround it in single quotes:

```r
# get the first 10 rows of the wingspan variable
bats %>%
  select('Wingspan (mm)') %>%
  head(10) # 10 rows
```

```
## # A tibble: 10 x 1
##    'Wingspan (mm)'
##    <chr>
## 1  23.5
## 2  na
## 3  31
## 4  52.5
## 5  40
## # ... with 5 more rows
```

 What is that %>% thing? It is known as a pipe. It sends the result of one operation to the next. So here `bats` is sent into `select` (get the variable you want), and the result of the `select` is sent into `head`, which, with the number 10, prints the top 10 rows. More about pipes in Section 6.1.

Ahh, now we can see the problem... some of the values of this variable are na and all the numbers have quotes around them; they would not if they were numeric. This is because the authors of the dataset used na as the indicator that a value was not available. We don't know why this value was not available, just that it was.

R can handle *not available* values just fine, but has to know the code we use to indicate then. Here R has failed to recognize an NA value, and has simply read it in as a word, and hence all other values in the variable also become words. It failed to recognize the missing value because read_csv(), by default, looks for NA, not na, as the missing-value entry. An efficient way to fix this is to do the original import with na specified as the NA value. Head back to your script and edit the data import code:

```
# read in the bat data, specifying the NA string as "na"
bats <- read_csv("data/bat_sex_diet_Mata_etal_2016.csv", na ="na")
```

Success. Now the wingspan variable: it is a col_double(). Excellent. This is how it should be: wingspan takes fractional values in this dataset.

> We just wrote 'Now the wingspan variable is a col_double()' but we didn't show this variable to you in the text. From now on, when you read something like 'look at the data' then please do this for yourself. For example, in the Console, type glimpse(bats) and hit Enter, or type View(bats) or click on bats in the Environment tab (as described above). You will need to do this a lot while working through this book.

3.6.1 HOW THE DATA ARE STORED IN R

When we imported the dataset into R it was put into an object termed a **tibble**, which itself is a kind of **data frame**. Both of these words ('tibble' and 'data frame') are the technical name of these objects. They are appropriate for data arranged in rows and columns, just like a spreadsheet such as that in Figure 3.2 and as previously described in Section 3.3. Keeping variables together in the same dataset is quite important, as keeping them separately would be quite error-prone, because for example, we might reorder one and not another, and thereby destroy their association. Therefore we keep related variables together in datasets.

To solidify these concepts, let's make a new script called 'MessingAbout.R' in the scripts folder. You can use this for the games/activities we show here that are not strictly part of the workflow on bat diets. You'll notice that you can have two scripts open at once as tabs!

At the top of this script, add a bit of annotation ('# just me messing about') and the `library(dplyr)` line and then use the following code to make a new tibble. First we create the variables:

```r
# just me messing about

# libraries
library(dplyr)

# make some variables for fun and insight into R coding
Person <- c("Owen", "Andrew", "Natalie", "Dylan")
Hobby <- c("Skiing", "Cycling", "Taxidermy", "Coding")
IQ <- c(1, 1, 100, 1)
```

The function `c` puts its arguments into a variable. For example, `Person <- c("Owen", "Andrew", "Natalie", "Dylan")` puts our names into a variable and then assigns that variable the name `Person`.

Now we can put the variables together in a tibble, using the `tibble` function:

```r
# make a tibble and explore it!
my_data <- tibble(Person, Hobby, IQ)
my_data
```

```
## # A tibble: 4 x 3
##    Person  Hobby            IQ
##    <chr>   <chr>         <dbl>
## 1 Owen    Skiing            1
## 2 Andrew  Cycling           1
## 3 Natalie Taxidermy       100
## 4 Dylan   Coding            1
```

Don't bother with row names. We can, if we like, name the rows. Don't bother, though… it's cumbersome to work with and row names are unnecessary. If we want to include some row-specific information, we can just add a variable/column containing it. (Situations in which one would want row names are beyond what we cover in this book, though are mentioned on the *Insights* companion website.[7]

Have a go in your messing-about script using the following code to look at our dataset, and figure out what it does:

```
# some R functions for looking at tibbles and data frames
head(my_data, n = 2)
tail(my_data, n = 2)
nrow(my_data)
ncol(my_data)
colnames(my_data)
View(my_data)
glimpse(my_data)
```

Tibble (tbl) objects. The **dplyr** and other tidyverse packages aim to make our life more pleasant. One way they do so is to use a special kind of data frame known as a tbl (pronounced 'tibble') that has some nice extra bells and whistles. When a *data frame* is shown in the Console, R will try to give every column and row, often resulting in a big mess. When a tibble is shown in the Console, it doesn't make a mess—it only prints what fits, and then summarizes the remainder.

3.7 Clean and tidy the data

In the previous section we checked that the data were imported into R correctly. Brilliant! The next step is to *clean* and *tidy* (if necessary) the data.

[7] http://insightsfromdata.io

Both are very important steps in the workflow, building and strengthening the foundation upon which the insights are built, and making for a more efficient, accurate, and reliable workflow for getting the insights. We've already identified some issues in this chapter and suggested quite a few in Chapter 1; we'll certainly find few more, and dealing with this will hopefully help you develop confidence to handle similar issues in your own data.

Recall that we are concerned now with experiencing the workflow, and less concerned with the details of how the R code works. In the next chapters we go into more detail about *how* the code is working, and about conceptual subjects. As we mentioned already, you may want to flick back and forth now, or make a list of things to look at later while focusing on the workflow. Do whichever seems to work best for you. If you can't decide, just read on… we're pretty sure things will work out well.

3.7.1 TIDYING THE DATA

We previously mentioned what tidy data are and why it is important to have tidy data (see Section 3.3). Fortunately, the bat diet dataset is supplied in tidy/long format (i.e. one observation per row, as we mentioned in Section 3.3), so we have no work to do in this respect. We will cover more details about tidying data (what to do when the data are not in this nice format) later, in Section 6.4. There are also examples of tidying some of the data in the food-diversity-polity Workflow Demonstration online.

3.7.2 CLEANING THE DATA

It's a good idea to review the kinds of things we might want to fix. Initially you might not see many, but with experience you'll start to see more! Here are some things we could think about doing, with the starred ones being what we will do with the bat dataset:

- *Refine variable names.
- *Format dates and times.
- Separate out any mixed information (there is none in the bat dataset).
- *Rename some values.

- *Check for and appropriately deal with any duplicate records.
- *Check for implausible and/or invalid values in variables.
- *Check for and appropriately deal with missing values.

Now, moving forward, let's shift back to the *bats_sex_diet.R* script. We'll start walking you through these starred changes. Remember, we've got a copy of the original data in R that we are working on, so none of this changes the Excel copy on the hard drive. But these changes are going to make it much easier to develop summary tables and visualizations that allow us to pursue the venerable insights on bats.

3.7.3 REFINE THE VARIABLE NAMES

Often we want to alter variable names from those in the original dataset. They might be non-intuitive. They might be long and therefore take a long time to type. In the case of the bat data, the variable names present many problems:

- Some of the names are non-intuitive: what does 'Sample' mean?
- Some of the names contain spaces.
- Some of the names contain brackets.

Let's first work with the non-intuitive names. Let us rename the Sample variable to something that more obviously shows that it contains the identity of individual bats:

```
# rename Sample to Bat_ID
# note that we are 'updating' or overwriting the R Copy of
# bats with the new version with a new variable name.
# we have not changed the Excel file!
bats <- bats %>%
  rename("Bat_ID" = "Sample")
```

The rename function from the **dplyr** package is relatively intuitive... we pipe to it the dataset to work with, and then do the renaming with "new_name" = "old_name".

We must be very careful when altering variable names, as mistakenly changing the meaning of a variable name will result in a data analysis *disaster*. So *be careful*. And, if possible, minimize alterations, and use string manipulation functions whenever possible (rather than renaming, as we did for the `Sample` variable, where we had no choice).

Now, let's work with the repeated variable name, `Order`. Recall the data contained two variables named `Order`. The first is the numbers 1:633 and could be useful for referencing a particular row of data. The other is the taxonomic order of the prey item, and it was automatically renamed to `Order_1` during data import. Let us rename the first to `Row_order` and the second to `Order`:

```
# rename Order to Row_order
# and Order_1 to Order.
bats <- bats %>%
  rename("Row_order" = "Order",
         "Order" = "Order_1")
```

Some of the variable names also have spaces and brackets. Although we can work with variable names with spaces, it's not so convenient. So, we prefer to alter the variable names to have no spaces. Because variable names are what are known as *strings* (more or less meaning they are words and not numbers), we will do some *string manipulation* to remove the spaces.

String manipulation is a common task when cleaning data, so we need to build experience and confidence with this. The **stringr** package is great for string manipulation, and is in the tidyverse. See Section 6.2 for more information about strings and how we can manipulate them.

Here is how we change the spaces in the variable names to underscores:

```
# change all spaces to underscores
names(bats) <- str_replace_all(names(bats), c(" " = "_"))
```

If you just tried to run the code above and got an error saying something like function str_replace_all not found, this is very likely because you didn't load the **stringr** package (though we asked you to do this above). And remember that to be able to load this package you must have installed it.

We use the function str_replace_all to do a replacement in all the strings of the first argument. The first argument is names(bats), i.e. the variable names of the bat dataset. The second argument is c(" " = "_"), which says replace spaces (note the space between the first two double quotes) with underscores. The names(bats) <- on the left-hand side replaces the variables names with the altered, new variable names. Done! And you can see it is impossible that we could mess up the meaning of a variable name with this string manipulation.

Brackets in variable names can also be a bit of a pain, so let's remove them using very similar code:

```
# get rid of brackets in names
names(bats) <- str_replace_all(names(bats), c("\\(" ="","\\)"=""))
```

Because R uses brackets in its code to know, for example to enclose the arguments of a function, we have to be careful when we refer to them in strings. The double backslash \\ before the brackets in the quotes allows R to know that we are asking for the text-string bracket, and not the functional bracket.

3.7.4 FIX THE DATES

We can also see that there is a Date variable. While it appears to be consistently entered, it's a good idea to ensure that dates and times are encoded properly. The **lubridate** package provides key functions to make this happen. (Further details about dealing with dates are in Section 6.3.) We can use a nice function from the **lubridate** package—dmy. We use dmy as, when we look at the Date variable, it seems clear that the day comes first, then the month, then the year. And yes, the function mdy() does the obvious, as do other orders of the three letters!

```
# use function dmy() from stringr package to
# encode Dates properly
bats <- bats %>%
  mutate(Date_proper = dmy(Date))
```

Here we use the mutate() function from the **dplyr** package to add a new variable called Date_proper. (See Section 5.1.3 for a detailed description of mutate.) Making a new variable allows us to compare the original date variable with the new one. All looks good (we do not get any errors, and the variable type is now Date—can you make sure of this?).

If we wish to clean up after this, we could now remove the original date variable, so we don't by accident try to use it:

```
# update our working copy of the data to
# remove the old Date column but leave the Date_proper one in
bats <- bats %>%
  select(-Date)
```

Here we use the select function from the **dplyr** package to remove the Date variable. The minus sign preceding Date *removes* that variable. See Section 5.1.1 for a detailed description of select.

3.7.5 RENAME SOME VALUES IN A VARIABLE

Many times, researchers use shorthand to code groups in a variable. We can see that's happened by looking at the entries in the Sex variable:

```
bats %>%
  select(Sex)
```

```
## # A tibble: 633 x 1
##    Sex
##    <chr>
## 1 F
## 2 F
## 3 F
```

```
## 4 F
## 5 F
## # ... with 628 more rows
```

We can guess that M means male and F means female, and we'd be correct. But for greater clarity, let's change them from codes to full words:

```
# use mutate and case_when
# for nested if-else like changes to values in a variable
# e.g. if Sex is M, make it Male....
bats <- bats %>%
  mutate(Sex = case_when(Sex == "M" ~ "Male",
                         Sex == "F" ~ "Female"))
```

We again do a mutate to create a new variable, though in this case our new variable is simply a writing over of the old version of Sex. The function case_when is nice for changing entries from one thing to another, particularly when there are more than two different entries to change. Each argument includes a squiggle, ~. On the left of it is the condition, and on the right is what to do if the condition is met. That is, we say to R 'if the value in the Sex variable is equal to M then please make the answer Male, thank you!' This may be familiar to some of you as a nested-if-else replacement.

We can do the same for the Age variable abbreviations:

```
# case_when replacement of Age values.
bats <- bats %>%
  mutate(Age = case_when(Age == "Ad" ~ "Adult",
                         Age == "Juv" ~ "Juvenile"))
```

We find it generally preferable to use complete words as values, and to not use codes and abbreviations. The result is greater clarity and less customization of tables and graphs, since the values in these are automatically taken from the dataset and therefore will also be complete words.

3.7.6 CHECK FOR DUPLICATES

Now that our variables are all named well, our dates are encoded, and the abbreviations used for some variables are changed, we can look for duplicates in the data. This is a standard safety check to perform on a dataset—whether it contains exactly duplicated rows, i.e. two or more rows with exactly the same entries.

This would often be a bit weird, we think. If there really were two separate measurements that happened to be exactly the same, then something must have differed between the two, but it has not been recorded in the dataset. Without that difference, we can't rule out that the observation may have accidentally been entered twice. We just don't know! So we think that, very generally speaking, it's worth looking for duplicated rows in datasets.

The function `duplicated` returns a list of TRUE/FALSE values, one value for each row. Any rows that are duplicates are given a value TRUE. Non-duplicated rows give FALSE. We can sum the number of TRUE values, and get zero if there are no duplicates:

```
# check for duplicate rows in the bats data
bats %>%
  duplicated() %>%
  sum()
```

```
## [1] 0
```

Good. No rows are exact duplicates. We can also check for duplicate observations with different measurements (i.e. we might not expect to have the same prey species recorded for the same bat on the same date). Every combination of Bat_ID and Sp._Nr. in the dataset should be unique:

```
# check for duplicates among specific combinations of variables
# select the ID, Sp._Nr. and Date_proper
bats %>%
  select(Bat_ID, Sp._Nr., Date_proper) %>%
  duplicated() %>%
  sum()
```

```
## [1] 0
```

Again, zero here means there are no duplicated combinations of Bat_ID and Sp._Nr. on any given date.

Checking for duplicates is very important. As you can see, we detect what we have to assume is inappropriate duplication in the food-diversity–polity Workflow Demonstration (and in other published datasets).

3.7.7 CHECK FOR IMPLAUSIBLE AND INVALID VALUES

We can also explore the data for implausible values of numeric variables, the number of unique entries in character variables, and so on. We can do this variable by variable like so:

```
# use summarise to calculate things about specific variables
bats %>%
  summarise(var_min = min(Wingspan_mm, na.rm = TRUE),
            var_max = max(Wingspan_mm, na.rm = TRUE))
```

```
## # A tibble: 1 x 2
##    var_min var_max
##      <dbl>   <dbl>
## 1      6.5    52.5
```

The minimum wingspan is 6.5 mm and the maximum is 52.5 mm (*Rhyacia simulans*, the dotted rustic, a moctuid moth).

We have used the summarise function for the first time. It is awesome—it allows us to calculate summary metrics (min, max, mean…) for specific variables and even by groups too. But we will leave explanation of it until later, as explaining now would break up the workflow experience. (If you want to learn about it now, look at Section 4.1 later in this Workflow Demonstration, or, to look at it in more depth, in Section 5.2.)

Our first insight from the data? You may have been wondering when our first insight from data would happen! They can creep up on us and not be noticed. Here could be our first… that the smallest prey species had a wingspan of 6.5 mm and the largest 52.5 mm. That is almost an order-of-magnitude difference in linear dimensions, which could equate to a three-orders-of magnitude difference in volume and mass (i.e. the prey species vary by 1000 times in size and perhaps energy content). Can you think of other simple insights that can come from using summarise() in this context of data checking? Perhaps the lowest- and highest-diversity diets recorded?

Here we can look at whether there are only males and females coded in the data. While there can be alternatives, in these data we don't expect any:

```
# use the dplyr function distinct() on the Sex variable to detect
# unexpected groups
bats %>%
  distinct(Sex)
```

```
## # A tibble: 2 x 1
##    Sex
##    <chr>
## 1 Female
## 2 Male
```

 You can use the usual UK (e.g. summarise, colour) or US (e.g. summa-rize, color) spellings while working in the **dplyr** and **ggplot** packages. Many other packages recognize only one or the other (US spelling being more common).

3.7.8 WHAT ABOUT THOSE NAS?

Since we're pretending we're the researcher, we should know how many NAs are in the dataset, and where they are. We will check this to make 100% sure. We already saw during data import that the Wingspan_mm variable contains some NAs. Let us get the number of NAs in the Wingspan_mm variable. We do this by using the summarise function to summarize the information in a variable, and tell it to give us the total number (sum) of values in Wingspan_mm that are NA (is.na(Wingspan_mm)):

```
# get sum of wingspan values that are NA
bats %>%
  summarise(num_nas = sum(is.na(Wingspan_mm)))
```

```
## # A tibble: 1 x 1
##   num_nas
##     <int>
## 1      78
```

We see that the `Wingspan_mm` variable contains 78 NAs. What about the other variables? We can use a special function and expression that we hope are logical, if not clear. The function is `across`, and it applies it applies a function *across* several variables, or even all of them:

```
# use the function across to sum the NAs in every column
# note the use of the function symbol "~" before sum() to make
# this work
# .cols specifies the columns and .fns defines the function(s)
bats %>%
  summarise(
    across(.cols = everything(),
           .fns = ~sum(is.na(.)))) %>%
  glimpse()
```

```
## Rows: 1
## Columns: 14
## $ Row_order   <int> 0
## $ Bat_ID      <int> 0
## $ Age         <int> 0
## $ Sex         <int> 0
## $ Sp._Nr.     <int> 0
## $ Species     <int> 0
## $ Class       <int> 0
## $ Order       <int> 0
## $ Family      <int> 0
## $ Pest        <int> 0
## $ Migratory   <int> 0
## $ Wingspan_mm <int> 78
## $ No._Reads   <int> 0
## $ Date_proper <int> 0
```

That's pretty cool. And, only the `Wingspan_mm` variable contains NAs. Which is good.

Now we have a decision to make. Are we going to remove the rows (i.e. observations) containing the NAs, so we don't have to deal with them further down the line, or leave them in and deal with them each time? There's no right or wrong here. Let's leave the NAs in for now, as it would be tricky to remove them without removing other quite valid and potentially useful information. This is because, to remove them, we would have to remove the entire row they appear in. More about NAs in a later chapter (in Section 8.4).

We've now got a lovely clean, tidy dataset that we can work with, so it's time to get some insights!

3.8 Stop that! Don't even think about it!

Before finishing up, we want to highlight a few things we think you should avoid doing. You might encounter these when taking other courses or talking with other R users. They are not wrong as such, but range from being unnecessary to being downright dangerous. For example, the `attach` function we mention in a moment can lead to unintended behaviours with no warnings or errors.

3.8.1 DON'T MESS WITH THE 'WORKING DIRECTORY'

If you have used R before, you may know that R is always 'looking' at a particular folder/directory on your computer. People call this the 'working directory'. You can find out the current working directory with the `getwd` function. People sometimes like to set the working directory in their script using the `setwd` function. This is a bad idea, because it means you can't easily move your project around without breaking it. You do not need to do this, and you should avoid doing it. Working with projects in the way we showed you is a much better solution. Whatever anyone says, don't use `setwd` for anything in this book (and try not to use it in your own work).

3.8.2 DON'T USE THE DATA IMPORT TOOL OR `file.choose`

It is possible to import data from a CSV file into R using the data import tool in RStudio. The steps are as follows:

1. Click on the Environment tab in the top right pane of RStudio.
2. Select `Import Dataset > From Text File (readr)...`
3. Select the CSV file to read into R, and click Open.
4. See if it looks OK in the dialogue box; mess with options if it doesn't.
5. Click OK and copy the code to your script.

We *do not* recommend this as a routine practice. This is mostly because the created code uses absolute and not relative paths. An implication is that if you move the project folder then the import code will stop working.

3.8.3 DON'T EVEN THINK ABOUT USING THE `attach` FUNCTION

Saying 'don't use `attach`' feels a bit like bringing to someone's attention a big red sparkly button with DO NOT PRESS written on it. They had no idea it was there and don't know what it does. Just like we don't know what would happen if we pressed the red sparkly button, we're not going to bother telling you what happens if you use `attach`. So this is a test of your trust in us. Don't do it. Just move on... nothing to see here.

3.8.4 AVOID USING SQUARE BRACKETS OR DOLLAR SIGNS

You will likely see R code with frequent dollar $ signs and square brackets []. We have old code with lots of them. This is fine. With the tidyverse approach, they are not necessary. So don't use them. At least, this is true while working with data and the methods in this book. If you start to use R for other programming tasks, e.g. running simulations, you may find these and other methods useful. You may also need to deal with them if you work with other people's code. We suggest that you learn about them when the need arises. If you feel that time is now, take a look at the relevant material on the *Insights* companion website.[8]

3.9 Summing up and looking forward

All that preparation is often not covered in data analysis classes, but it is clearly essential for working with your own data. You've organized your computer, put data into specific locations to make access easy, and developed a script that deals with many issues surrounding variable names, values in variables, dates, NAs, data ranges, and more. Along the way, you've also obtained some insights, like the smallest and largest prey species and which bat species ate them. You may have acquired others too!

These steps and functions are a recipe for starting to manage data in projects. They set you up for subsequent use of functions that are highly geared towards generating insight via tables and graphs. Without such a solid foundation, anything we build on it may be suspect. Furthermore, taking care of foundations should increase your confidence, and also your knowledge of your data. Both of these are very useful for the next steps… getting insights.

To recap the highlights of the first part of the workflow:

- Carefully specify the question to be answered.
- Be clear about the response variables.
- Design and perform the study.
- Get the data, and import them into R.
- Check the data we've imported are what we expected.
- Clean and tidy the data (if necessary).
- Gain some basic insights about specific variables along the way.

---

[8] http://insightsfromdata.io

From R, we've introduced you to functions from several packages:

- Reading data in with the **readr** package.
- Several functions in the **dplyr** package for data manipulation and summarization.
- Some string manipulation with the **stringr** package.
- Some cleaning of dates using the **lubridate** package.

Within these packages, we've used many functions: `library`, `read_csv`, `View`, `slice`, `tibble`, `head`, `tail`, `nrow`, `ncol`, `colnames`, `rename`, `str_replace_all`, `names`, `mutate`, `filter`, `select`, `sum`, `duplicated`, `summarise`, `summarise_all`, `distinct`, `is.na`.

We strongly advise you to keep a list of the functions you are using, what they were used for, what were their arguments, and which package they are in. Perhaps get a little notebook to keep such notes in, or build a table in a spreadsheet. Or print one of the relevant RStudio cheat sheets and circle the functions you learned how to use. Not only will these notes be a nice reference for you, they will also show you how much progress you're making.

The *Insights* companion website[9] provides all the R scripts from this chapter and the next (part 2 of the workflow) in one place, with comments made as if we were the researcher, but with little explanation or narrative. This is what our script would look like if we were just working through the workflow (rather than also guiding your learning).

Another useful thing to do is to make a list of the things in this section that you did not fully understand. It might be quite a long list, as the purpose of this chapter was not for you to develop an understanding of the details. While going through the remainder of this book you will, we hope very much, be able to put a tick against all these things to indicate you do then understand them.

Now let's move on to the next part of the workflow, and get some more insights from the bat diet dataset!

---

[9] http://insightsfromdata.io

Workflow Demonstration part 2

Getting insights

W e're ready to make our first insights… this would be a pretty exciting time for any researcher. The empirical work is done, the data are collected, typed in, checked, tidied, cleaned, and now it's the first time to get some answers.

We think it's pretty important to remind you that, as a researcher or data scientist, at this stage it might be a good idea to grab a pen/pencil and paper and make an outline of what you want to do. This might even include sketches/cartoons of graphs and tables, with labels, that you want to make. Moreover, these sketches might even include patterns you might have been expecting, given your background with the questions that led to the data collection.

Now, before we dive in, and to ensure you know exactly where in the workflow we are, please print out a copy of the workflow, either the diagram or the checklist, both of which are available on the *Insights* companion website.[1] Then first recap part 1 of the workflow, perhaps writing some notes for each step, and perhaps noting the R functions used in each step

---

[1] http://insightsfromdata.io

Insights from data with R: An Introduction for the Life and Environmental Sciences. Owen L. Petchey, Andrew P. Beckerman, Natalie Cooper and Dylan Z. Childs, Oxford University Press (2021). © Owen L. Petchey, Andrew P. Beckerman, Natalie Cooper and Dylan Z. Childs.
DOI: 10.1093/oso/9780198849810.003.0004

(though of course many functions are useful in many steps). Then preview what is coming in this second part of the workflow: getting the insights that are the answers to our original questions.

4.1 Initial insights 1: Numbers and counting

We suggest starting with some basic insights. We suggest focusing on further questions about specific variables. For example:

- How many bats were caught?
- How many males, and how many females?
- How many adults, and how many juveniles?
- How many species of prey were found across all the poops analysed?
- What is the distribution of prey sizes?

Some of these are simple in that they are about a single variable, like the number of bats. Others are slightly more complex, like evaluating the number of males and females, which are the two groups in the sex variable. Regardless, these are what we call 'safety-checking insights'. They are the types of insight that you might already know, because you were responsible for collecting the data. Or they are things you really should know, if you were not responsible for collecting the data.

To gain these insights, we will (re)introduce a wide variety of very nice functions in the **dplyr** package and give a brief account of how they work. There are more in-depth explanations later (see Section 5.2.2).

4.1.1 OUR FIRST INSIGHTS: THE NUMBER, SEX, AND AGE OF BATS

Here is the code to get the total number of bats:

```
# how many distinct bats are there...
# there are multiple observations of each bat
# n_distinct deals with this
bats %>%
  summarise(n_distinct(Bat_ID))
```

```
## # A tibble: 1 x 1
##    `n_distinct(Bat_ID)`
##                   <int>
## 1                   143
```

In normal words, we ask for the number of distinct values in the `Bat_ID` variable. The first line, `bats  %>%`, 'passes' (i.e. 'pipes') the `bats` data frame to the `summarise` function in the next line.

Inside the `summarise` function, we specify a new function `n_distinct` and tell it we want it to work on the `Bat_ID` variable. We must ask for the number of distinct values, because each bat ID can occur multiple times, one for each of the different prey species found in the sample of its poop.

The answer we get is that all our hard work in the field resulted in us catching and getting the poop of 143 bats. Awesome! We love bats, and we got 143 bat poops!

The answer is provided in a tibble (see Chapter 3). The name of the column returned by `summarise` is `n_distinct(Bat_ID)`, which is a bit ugly. We can instead modify the code to give the column a name we like (e.g. `num_bat_IDs`) like this:

```
# how many distinct bats are there...
# there are multiple observations of each bat
# n_distinct deals with this
bats %>%
  summarise(num_bat_IDs = n_distinct(Bat_ID))
```

```
## # A tibble: 1 x 1
##    num_bat_IDs
##          <int>
## 1          143
```

Now let's find out how many female bats and male bats we caught. To do this, we employ the `group_by` function from **dplyr**:

```
# create groups by Sex - Male and Female
# ask for distinct observations in each group
bats %>%
  group_by(Sex) %>%
  summarise(num_bat_IDs = n_distinct(Bat_ID))
```

```
## `summarise()` ungrouping output (override with `.groups`
argument)

## # A tibble: 2 x 2
##    Sex      num_bat_IDs
##    <chr>          <int>
## 1 Female            69
## 2 Male              74
```

The first and third lines of the command are the same as before. We have added the line group_by(Sex) to tell summarise to do its work separately for each sex. The counts exactly match those in the abstract of the paper! The answer comes back in a dataset (tibble) with two rows and two columns ('A tibble: 2 x 2'). The first column is the sex, and the second is the number of bats. We see that we caught 69 female bats and 74 male. It is quite nice that we caught very similar numbers of each sex.

Let's do the same for juvenile and adult bats. We only need to change the argument in group_by from Sex to Age:

```
# create groups by Age - Juvenile and Adult
# ask for distinct observations in each group
bats %>%
  group_by(Age) %>%
  summarise(num_bat_IDs = n_distinct(Bat_ID))
```

```
## `summarise()` ungrouping output (override with `.groups`
argument)

## # A tibble: 2 x 2
##    Age      num_bat_IDs
##    <chr>          <int>
## 1 Adult            102
## 2 Juvenile          41
```

More adults than juveniles were caught. Now, what about the number of juvenile male, juvenile female, adult male, and adult female bats? Can you guess how we do this? Yes! You can specify both Sex and Age as arguments in the group_by function:

```
# summary by two grouping variables
bats %>%
  group_by(Age, Sex) %>%
  summarise(num_bat_IDs = n_distinct(Bat_ID))
```

```
## `summarise()` regrouping output by 'Age' (override with
`.groups` argument)

## # A tibble: 4 x 3
## # Groups:    Age [2]
##    Age        Sex      num_bat_IDs
##    <chr>      <chr>          <int>
## 1 Adult      Female            55
## 2 Adult      Male              47
## 3 Juvenile   Female            14
## 4 Juvenile   Male              27
```

We get a data table with four rows (one for each combination of age and sex) and three columns. The values in the first two columns tell us the age and sex that the entries in the third column refer to, i.e. we caught 55 adult female bats and 47 adult male ones.

Fantastic! We've got several types of insights here about the bats, both overall and in groups. We've not yet gained insight into how many species of prey were found across all the poops analysed, or the distribution of prey sizes, but this is coming.

You just experienced two of the most amazing R functions: group_by and summarise. (You also experienced more of the magical pipe, %>%). These and other functions in the **dplyr** package have, in the last few years, revolutionized how people do data analysis in R. They make generating summary tables for insight very easy and intuitive, we think. Go back and read the code a few times… for example, start with the bat data, pass it to group_by to divide it up by sex and age, and then use summarise to calculate the numbers of unique bat individuals in each group.

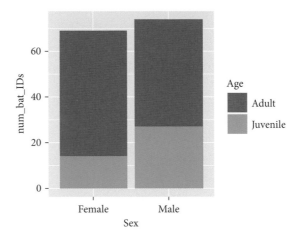

Figure 4.1 A first insight! Numbers of male and female bats of each age.

Don't worry if you don't yet feel 100% comfortable in your under-standing of how `summarise` or `group_by` works. The point of this chapter is for you to experience and see the general workflow. We will explain again and in more detail in Section 5.2.1.

Before we do that, let's translate some of this tabular insight into graph-ical insight by making a graph showing the number of bats by sex and age with the following code (the resulting graph is in Figure 4.1):

```
# make summary data, and give it a name to use
bats_Age_Sex <- bats %>%
  group_by(Age, Sex) %>%
  summarise(num_bat_IDs = n_distinct(Bat_ID))
```

```
## `summarise()` regrouping output by 'Age' (override with
`.groups` argument)
```

```
# basic use of ggplot
bats_Age_Sex %>%
  ggplot() +
  geom_col(mapping = aes(x=Sex, y=num_bat_IDs, fill=Age))
```

Here is your first sight of how we make all the graphs in this book: with ggplot and associated functions (e.g. geom_col). These come from the amazing **ggplot2** package, which has become very popular over the last few years because it is, well, fantastic! The following description of how ggplot works is very brief and maybe not perfectly clear. Fear not, all will be revealed in Chapter 7, a deeper dive into ggplot.

To start, we made sure that the summary table we made was assigned to a name, bats_Age_Sex. This means we can now use this mini-data frame (it should be visible in your Environment pane). To get the graph we used two lines of R, one starting with ggplot and the other with geom_col. We piped ggplot the dataset containing the data to be plotted. In the next line we ask for a plot with columns (bars) by using the geom_col function. We give geom_col the 'aesthetic mappings', specifically aes(x=Sex, y=num_bat_IDs, fill=Age) (note that aes is itself a function). (As we said, that will likely not make perfect sense . . . but all will be revealed later. And if you can't wait, skip forward to Chapter 7.)

Note that we *add* geom_col to ggplot—there is a plus sign at the end of the ggplot line. This builds, on top of the axes, the bars at the correct height, with colours coded by age, and a handy legend produced at the side.

Now we have two forms of insight—tables and graphs—about one aspect of these data—the number of bats we found. Your challenge now is to modify the code we just used to estimate the number of distinct prey items (diversity) found among bat poops in each age and sex combination. Can you make the table and the graph?

4.2 Initial insights 2: Distributions

We now know how many bats of each type we caught. And, hopefully, you did the exercises and so know the number of prey species of each type detected in the poops. Let's move on to a very important type of insight about a variable: its distribution. Recall that one of our research questions concerns the size of the prey species, so let us look at the distribution of

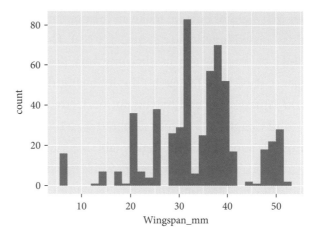

Figure 4.2 Frequency distribution of the size (wingspan) of the prey.

the variable `Wingspan_mm` (this is what we use as a measure of the size of the prey).

Looking at frequency distributions is very important because they show the 'shape' of the *sample distribution*, and that shape is very important. In Section 8.2 we detail what is a *sample distribution*, and in Section 8.3.2 we go into details of how to summarize features of a sample distribution (like the mean).

Here is the script to plot the frequency distribution of the `Wingspan_mm` variable (the result is in Figure 4.2):

```
bats %>%
  ggplot() +
  geom_histogram(mapping = aes(x = Wingspan_mm),
                 bins = 30)
```

```
## Warning: Removed 78 rows containing non-finite values
## (stat_bin).
```

As before, we pipe the data into the `ggplot` function. Then we add to the `ggplot` function a histogram with the `geom_histogram` function. We only specify the x-variable, `Wingspan_mm` because a histogram's y-axis is made up of counts of observations of the x-variable in 'bins' or subsets of this x-variable. We then ask for the `Wingspan_mm` variable to be represented in 30 equally sized bins; `geom_histogram` does the counting for you! The result is that each bar is the count of the number of values of `Wingspan_mm` that fall within the *bin* of that bar.

OK, great, but what's a bin? Think of people fishing on the deck of a boat sorting fishes into three sizes, small (less than 1 kg), medium (1-2 kg), and large (2-3 kg). And that there is a bin/bucket for each size class. The count is the number of fish in each bin/bucket.

Try changing the number of bins. See what effect there is on the shape of the histogram. Do you see what's happening to the bin width as you change the number of bins?

When you made the graph, you got another message that said `Warning message: Removed 78 rows containing non-finite values (stat_bin)`. Warnings mean that R has done what we asked but wants to alert us to a possible issue. If we expected all the data to be plottable, this might be cause for concern. But thanks to all that data checking we did earlier in the workflow, we know there are NAs in the wingspan data, 78 to be precise. So this is just telling us R has not plotted the NAs.

4.2.1 INSIGHTS.... YOU'VE DONE IT!

Now to the important questions—the insights from this exercise. What are the features and shape of this distribution? First, it has a lower limit of zero... wingspan cannot be negative, and it has an upper value, which we've calculated in the last chapter too! Second, it is not symmetric, though the peak is somewhere near the middle. We can summarize the shape of this (and any distribution) with a few numbers. The mean (the average value), the median (the middle value when they are sorted from smallest to largest), and the mode (the most common value) are measures of where the centre of the distribution is located. The spread of the distribution can be measured by its variance and standard deviation. And the symmetry by the skewness. Section 8.3.2 goes into more detail about these types of summaries of distributions.

Look at the distribution of wingspans (Figure 4.2) and write down guesses for the mean and median, and guess whether the mean or median is larger.

Now, let's use the power of R and **dplyr** to get more accuracy around these guesses. Remember that summarise can deliver more than one summary metric:

```
# generate in an insightful table of calculations
# the mean, median and standard deviation
# of the wingspan_mm variable
bats %>%
  summarise(mean_wingspan = mean(Wingspan_mm),
            median_wingspan = median(Wingspan_mm),
            sd_wingspan = sd(Wingspan_mm))
```

```
## # A tibble: 1 x 3
##    mean_wingspan median_wingspan sd_wingspan
##            <dbl>           <dbl>       <dbl>
## 1             NA              NA          NA
```

Huh, hang on, why do we get NAs? Ah, it's those pesky NAs in the wingspan again, making it impossible to get a mean without telling R to ignore them. This is easy to do using na.rm = TRUE (rm = 'remove'):

```
# calculate in an insightful table
# the mean, media and standard deviation
# of the wingspan_mm variable, dealing with NA values!
bats %>%
  summarise(mean_wingspan = mean(Wingspan_mm, na.rm = TRUE),
            median_wingspan = median(Wingspan_mm, na.rm = TRUE),
            sd_wingspan = sd(Wingspan_mm, na.rm = TRUE))
```

```
## # A tibble: 1 x 3
##    mean_wingspan median_wingspan sd_wingspan
##            <dbl>           <dbl>       <dbl>
## 1           33.6              35        9.59
```

Remember that the median is the value of the wingspan at which 50% of the data are below it and 50% are above it. The mean is the average wingspan. And the standard deviation captures an estimate of how many units of wingspan above and below the mean together capture about two thirds of the distribution. We see that the mean is lower than the median, which results from a slight negative/left skew, i.e. there are fewer small values than large values of the wingspan.

Why is all this detail so important? Well, lots of statistical analyses are based on assuming a particular shape of the distribution (e.g. normal or not), and knowing the values of the mean and standard deviation. So we need to be able to recognize the shapes and their features and even calculate them.

We just asked you to make an informed guess of the mean and median and write them down. We suggest you make informed guesses like this whenever you can. Comparison of our guess with what R then tells us is a great *check and balance* between our understanding of the data and the understanding revealed by our R code.

In the detail of the code and discussion above, we seem to actually have missed out the insights! There are some juicy ones:

- Bats eat a range of prey with different wingspans.
- The average wingspan of these prey is 33.65 mm.
- Overall, bats tend to eat more larger items than smaller ones.

These may not seem the most groundbreaking of insights, but they are important features of the data that we were previously unaware of. Give yourself a pat on the back... install the `praise` add-on package and run the following:

```
library(praise)
praise()
```

To go ahead and make some more insights, we need to do a little more of the workflow and create some new variables for our final analyses.

4.3 Transform the data

In the previous sections we made some basic insights about the bat diet data, but we have not yet gained insight into our core questions: are there differences in number of prey species, average size of prey species, and proportion of migratory prey between male and female bats? You have probably noticed that none of these three variables exist in the dataset. Let us go through the process now of creating these variables so that we can gain our insights.

The three response variables we need are as follows.

Number of prey species eaten by each bat. Recall that the dataset contains one row for each prey species eaten by each bat. Hence, if a bat eats three prey items, it should have three rows. So we just need to count the number of rows for each bat individual. Put another way, we want to group the data by bat individual identity (contained in the variable `Bat_ID`), and then count the number of rows in each group. Here is how we do that:

```
prey_stats <- bats %>%
  group_by(Bat_ID) %>%
  summarise(num_prey = n())
```

```
## `summarise()` ungrouping output (override with `.groups`
argument)
```

The first line pipes the dataset `bats` into the second line. The second line uses the `group_by` function to define the variable that contains the groups. The third line says 'please `summarise` the data by counting the number of rows in each group' (it uses the `n()` function to do this) and 'put the values in a variable called `num_prey`'. Take a look at the result:

```
prey_stats
```

```
## # A tibble: 143 x 2
##    Bat_ID num_prey
##     <dbl>    <int>
```

```
## 1      366          2
## 2      367          9
## 3      584          6
## 4      598          5
## 5      606          6
## # ... with 138 more rows
```

A tibble with 143 rows, one for each bat (great, as we know there are 143 bats). The first column is the bat identity and the second is the number of prey eaten by each bat. Super! We did it. But before we move on to make the other two response variables, let's finish this one properly.

First, a safety check (one of those *checks and balances* that contribute to accurate and robust insights). We have just made an assumption that the same prey species is never recorded twice for the same bat. We will check if this is so by calculating the number of prey species eaten by a second method: getting the number of distinct prey identities for each bat. Note that we use exactly the same code as before, but just add another line in the summarise function. This line asks for the number of distinct (n_distinct) prey species (Sp._Nr.) for each group (bat identity). The result is saved in the variable num_prey1:

```
prey_stats <- bats %>%
  group_by(Bat_ID) %>%
  summarise(num_prey = n(),
            num_prey1 = n_distinct(Sp._Nr.))
```

```
## `summarise()` ungrouping output (override with `.groups`
argument)
```

If you look at this dataset you will see that num_prey and num_prey1 contain the same values, so our check shows that we correctly understood the dataset and calculated the number of prey species. Great!

A final thing before moving on to the other response variables—you may have noticed that our new dataset has no record of the age or sex of each bat. But we need this to answer our questions: we want to look at how prey number varies with bat sex and age, so we need to keep those variables in the dataset. We can keep them by adding them to the group_by function:

```
prey_stats <- bats %>%
  group_by(Bat_ID, Sex, Age) %>%
  summarise(num_prey = n(),
            num_prey1 = n_distinct(Sp._Nr.))
```

```
## 'summarise()' regrouping output by 'Bat_ID', 'Sex'
(override with '.groups' argument)
```

Sweet! Now let's get the *mean wingspan of the prey eaten by each bat*, and the *proportion of migratory prey*. To do this, we just add two more calculations to the existing summarise function:

```
prey_stats <- bats %>%
  group_by(Bat_ID, Sex, Age) %>%
  summarise(num_prey = n(),
            num_prey1 = n_distinct(Sp._Nr.),
            mean_wingspan = mean(Wingspan_mm, na.rm = TRUE),
            prop_migratory = sum(Migratory == "yes") / n())
```

```
## 'summarise()' regrouping output by 'Bat_ID', 'Sex'
(override with '.groups' argument)
```

The line with mean_wingspan = mean(Wingspan_mm, na.rm = TRUE) should be pretty self-explanatory. The line with prop_migratory = sum(Migratory == "yes") / n() is a bit more involved. It contains the logical operator ==, which asks if the left- and right-hand sides are equal. If they are then it gives a TRUE, and if not it gives a FALSE. The sum adds up the TRUEs (each of which is worth 1) and the FALSEs (each of which is worth 0). So for each bat it counts the number of prey species that are migratory, and then divides by the total (divide by n()). Magic, eh?! Not really, but at least pretty elegant. We now have the three new variables and we can move on to answering our three main questions.

Logical operators like == are operators just like + and – but their answer is yes or no (hence their name, logical operators). They include > (greater than), >= (greater than or equal to), < (less than), <= (less than or equal to), ! = (not equal to), & (and), and | (or). All are used like questions asking something about what appears on their left and right sides. We'll see more examples throughout the book and then explain them in more detail.

Note that several times above we assigned the result to the object `prey_stats`. Hence, each time we destroyed the previous version of `prey_stats`. This is not just OK, but actually a good idea. Keeping variables in the same data frame whenever possible, and not creating multiple data frames when we need only one, is good for safety and efficiency.

From here on in this chapter we work with the `prey_stats` data, so please make sure you fully understand how they were obtained.

4.4 Insights about our questions

Now we're ready to make our proper insights of the kind that we could add to our reports/papers/thesis chapters. We'll focus first on exploring the distributions of the three response variables, and then progress towards answering our three biological questions.

4.4.1 DISTRIBUTION OF NUMBER OF PREY

We will examine the shape of the *sample distribution* of the data, i.e. the shape of the frequency distribution, also known as the shape of the histogram. In Section 8.2 we detail what is a sample distribution, and in Section 8.3.2 we go into details of how to summarize features of a sample distribution (like the mean). Hence, as you read on in this chapter, focus on the workflow and the meaning of the data in the context of the question being answered.

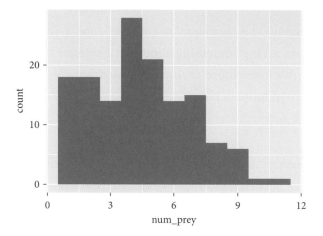

Figure 4.3 Frequency distribution of number of prey in bats' diets, with number of bins specified.

Let's look at the distribution of the number of prey, setting the bins as we learned previously, with the resulting graph in Figure 4.3:

```
prey_stats %>%
  ggplot() +
  geom_histogram(mapping = aes(x = num_prey),
                 bins = 11)
```

Now to the important question. What shape is this distribution? First, it has a lower limit of zero... the number of prey eaten by a bat cannot be negative. Second, it is not symmetric, though the peak is somewhere near the middle. Section 8.3.2 gives more detail about these types of summaries of distributions.

Look at the distribution in Figure 4.3 and write down guesses for the mean and median, and thus guess whether the mean or median is larger. The answers are obtained with this code:

```
prey_stats %>%
  # we made prey_stats with grouping variables
  # we need to ignore this meta-information to get the mean
    and median
  ungroup() %>%
```

```
summarise(mean_num_prey = mean(num_prey, na.rm = TRUE),
          median_num_prey = median(num_prey, na.rm = TRUE))
```

```
## # A tibble: 1 x 2
##   mean_num_prey median_num_prey
##           <dbl>           <int>
## 1          4.43               4
```

We see that the mean is greater than the median, which results from positive/right skew. That is, there are a few rather large values, and quite a lot of small values.

Why is all this so important? Well, much of statistics is based on assuming a particular distribution shape, so we need to be able to recognize the shapes and their features. Though we're not going into it further, the Poisson distribution is often a very good starting point when one wants to do statistics on count data.

4.4.2 SHAPES: MEAN WINGSPAN

Let's do the same for the mean wingspan of eaten prey species. The following code makes Figure 4.4 and gives the mean and median:

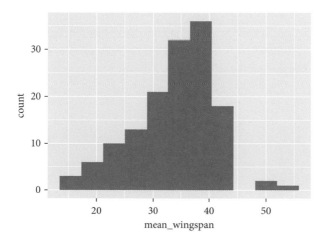

Figure 4.4 Frequency distribution of mean wingspan of prey in bats' diets.

```
prey_stats %>%
  ggplot() +
    geom_histogram(mapping = aes(x = mean_wingspan),
                   bins = 11)
```

```
prey_stats %>%
  ungroup() %>%
    summarise(mean_mean_wingspan = mean(mean_wingspan, na.rm = TRUE),
      median_mean_wingspan = median(mean_wingspan, na.rm = TRUE))
```

```
## # A tibble: 1 x 2
##    mean_mean_wingspan median_mean_wingspan
##                 <dbl>                <dbl>
## 1                33.7                 34.5
```

The distribution is also rather asymmetric, and we see the mean is less than the median. The values are continuous (i.e. do not have to be whole numbers).

4.4.3 SHAPES: PROPORTION MIGRATORY

And now the same for the last variable, the proportion of migratory species in the diet. The following code makes Figure 4.5 and gives the mean and median:

```
prey_stats %>%
  ggplot() +
  geom_histogram(mapping = aes(x = prop_migratory),
                 bins = 10)
```

```
prey_stats %>%
 ungroup()%>%
  summarise(mean_prop_migratory = mean(prop_migratory, na.rm = TRUE),
     median_prop_migratory = median(prop_migratory, na.rm = TRUE))
```

```
## # A tibble: 1 x 2
##    mean_prop_migratory median_prop_migratory
##                  <dbl>                 <dbl>
## 1                0.452                 0.429
```

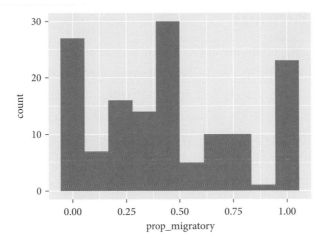

Figure 4.5 Frequency distribution of proportion of migratory prey in bats' diets.

A very different distribution again, as you might have guessed. The minimum is zero and the maximum is one. Here the mean and median are relatively similar. And there isn't one obvious peak (i.e. most common value)… there might be two or three, depending on how many bins we use. When doing statistics on this kind of data, we often start with the binomial distribution.

In addition to the insights provided by examining these histograms, we can again check that the data are what we expect. For example, are there wildly small or large values, or impossible values (e.g. negative numbers of prey or proportions larger than one)? And, as mentioned, looking at the shapes of distributions is a very good prelude to doing statistics.

We won't say a great deal more about theoretical distributions such as the normal, Poisson, and binomial distributions. You would learn much more about these in a statistics course. It's enough now to see the different types of data they describe, and to be happy that you've experienced three of the most important and common theoretical distributions in statistics!

4.4.4 RELATIONSHIPS

Getting insights often involves looking for relationships. If we have a manipulated variable in a well-designed experiment, a relationship/association can indicate a causal effect. In a study without a manipulated variable, such as this bat diet study, we cannot be sure that any relationship we find is the result of causation. As they say, 'correlation does not imply causation', or, perhaps more appropriately, 'correlation implies some unknown pathway of causation connecting two variables'.

Anyway, here let's focus on looking at relationships that will answer our three main questions.

Dietary sex differences

First let's look at the relationship between bat sex and prey size with this code, which produces the graph in Figure 4.6:

```
prey_stats %>%
  ggplot() +
  geom_beeswarm(mapping = aes(x = Sex, y = mean_wingspan))
```

We again use `ggplot` to make the graph, and add `geom_beeswarm` (which nicely distributes data points to stop them obscuring each other),

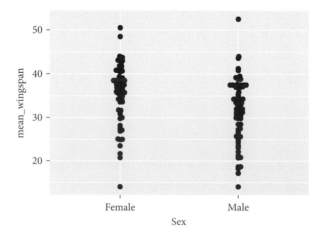

Figure 4.6 Difference in size of prey in the diets of male and female bats.

giving it two aesthetic mappings, one for the *x*-axis (x = Sex) and one for the *y*-axis (y = mean_wingspan). (Recall that Chapter 7 provides a deeper dive into ggplot).

Look at Figure 4.6 and guess the mean values of mean_wingspan of male bats and of female bats, and the difference between the two means. Perhaps you agree that it looks like the female bats eat prey that are about 5 mm larger in wingspan than the prey that the males eat. Let's check if this guess is anywhere near what the data really show:

```
prey_stats %>%
  group_by(Sex) %>%
  summarise(mean(mean_wingspan, na.rm = TRUE))
```

```
## `summarise()` ungrouping output (override with `.groups`
argument)
```

```
## # A tibble: 2 x 2
##    Sex     `mean(mean_wingspan, na.rm = TRUE)`
##    <chr>                                 <dbl>
## 1 Female                                 35.6
## 2 Male                                   31.9
```

Our guess of a difference of about 5 mm was not so far from the actual difference in this data. That is, we can see a lot from the data.

We have seen that the difference between the means is about 4 mm, but is this a lot or not a lot? One way to figure that out is to compare this difference with the amount of difference within the females and within the males. That question is similar to asking if we'd be likely to find the observed difference if there were no real difference. One can answer that question by randomizing the data and seeing if the real pattern is easily distinguished from a randomized dataset. (Another way to answer the question is via a statistical test, but we're not going there.)

```
prey_stats %>%
  ggplot() +
  geom_beeswarm(mapping = aes(x = Sex, y = sample(mean_wingspan)))
```

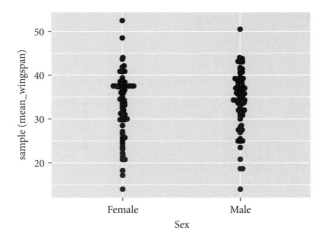

Figure 4.7 Randomized size of prey in the diets of male and female bats.

This code includes `sample(mean_wingspan)`, which randomizes the order of the plotted `mean_wingspans`, i.e. it shuffles them randomly between male and female bats. The code gives Figure 4.7. Judge for yourself if you think this looks very different from the real data. And make the graphs lots of times and record how many times you think the sex difference in the randomized data looks the same as the sex difference in the real data. If you get a very small number, you are judging that the real pattern is quite unlikely to have arisen by chance alone, i.e. there really seems to be a difference!

Let's now look at whether females or males consume more migratory prey. The code below produces Figure 4.8, which on first sight may look a bit weird. There is at least a hint that females consume a greater proportion of migratory prey, but the difference is not so massive compared with the variation within the sexes:

```
prey_stats %>%
  ggplot() +
  geom_beeswarm(mapping = aes(x = Sex, y = prop_migratory))
```

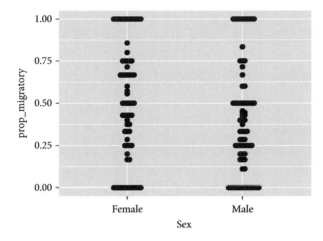

Figure 4.8 Difference in migratory nature of prey in the diets of male and female bats.

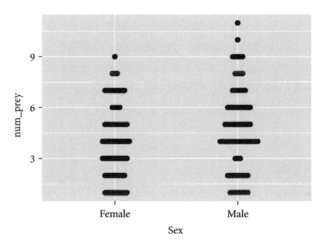

Figure 4.9 Difference in number of prey in the diets of male and female bats.

And, finally, let's look at the number of prey with this code, which produces the graph in Figure 4.9. There's at least a hint that females eat fewer prey species than males, but the difference is not massive:

```
prey_stats %>%
  ggplot() +
  geom_beeswarm(mapping = aes(x = Sex, y = num_prey))
```

Notice that we choose to plot the data points, and not bar charts or box-and-whisker plots. We believe it is better to plot the data points rather than summaries of them.

Age–sex interactions

Now let us concern ourselves with interactions. Often we are very interested in whether effects are context dependent—e.g. whether the effect of sex depends on age. Looking for such 'interactions' can produce really quite nice insights.

Focusing on prey size, we ask if the difference between the sexes depends on age. For this, we need to also display which data points are for which age. We will do this by creating two graphs (i.e. two *facets*), one for each age class (Figure 4.10), using the function `facet_wrap(~ Age)` addition to `ggplot`. We ask for facets, and use the squiggle ~ followed by the variable that contains the groups to be used in each facet. (You're probably wondering why the squiggle. Good. But the answer is not important here. Trust. Us. It's a need-to-know issue at this point, and you don't *need* to know.)

```
prey_stats %>%
  ggplot() +
  geom_beeswarm(mapping = aes(x = Sex, y = mean_wingspan)) +
  facet_wrap(~ Age)
```

Now, look carefully at Figure 4.10 (to which we have also added the mean of each sex–age combination as a blue circle). On each facet, imagine a line joining the two means. Do you think the two lines are approximately parallel? If so, you see a similar difference between males and females regardless of whether they are adults or juveniles. We don't see that the angles/slopes of the lines are very different, and so don't see a difference in the sex–size relationship between the two ages:

```
## `summarise()` regrouping output by 'Sex' (override with
`.groups` argument)
```

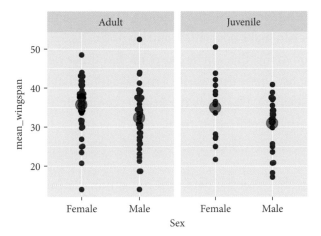

Figure 4.10 Difference in prey size in the diets of male and female bats of two ages.

4.4.5 COMMUNICATION (BEAUTIFYING THE GRAPHS)

Because this is the first time (and one of the few times) we dip into 'beau-tification', let's go over our philosophy: only change what needs changing; don't change it if it ain't broke. Put another way, let's agree to only spend our valuable time changing the things that *need* to be changed, and let's keep things that might be *nice* to change for a rainy day when we've nothing better to do than fiddle with graphs, or when we have some important procrastination that can be accomplished by fiddling with graphs.

How do we know what *needs* to be done to beautify a graph? Show your graph to a friend and ask them what they see. The aim is that they can see the answer/insight without your help or any accompanying text. If they can't figure it out, then help them to do so, and ask what should be changed to make it clear. They probably won't suggest changing the font, colour of background, or other things like that.

To be clear, we could change more or less anything we like: colours, back-grounds, grids, axis tick mark positions, axis tick labels, label orientations, and many, many other such things. But why? It might be that you actually want to make a fancy, colourful, popping figure for a public presentation or paper. You can make one with R and `ggplot`.

We mentioned that this will be one of the few times we dip into 'beautification' of graphs. This is because it's a massive topic that has been very well dealt with elsewhere. Please see the *Insights* companion website[2] for good sources to learn 'beautification' from.

If at all possible, plot the data, and not summaries of the data. That is, plot the individual data points. This is well aligned with the *Nature* publishing group's instructions: 'Present data in a format that shows data distribution (dot-plots or box-and-whisker plots). If using bar graphs, overlay the corresponding dot plots. Confirm that all data presentation meets these requirements and that individual data points are shown.'

4.4.6 BEAUTIFYING THE WINGSPAN, AGE AND SEX GRAPH

When we looked at Figure 4.10 we could see only one thing we definitely need to change: the label of the *y*-axis. This is quite straightforward to do by adding a `ylab` to `ggplot`; the result appears in Figure 4.11:

```
prey_stats %>%
  ggplot() +
  geom_beeswarm(mapping = aes(x = Sex, y = mean_wingspan)) +
  facet_wrap(~ Age) +
  ylab("Mean wingspan of prey items\neaten by individual bats")
```

Recall when we changed the entries in the `Sex` variable from `M` to `Male` and `F` to `Female`, and also changed the entries in the `Age` variable to `Adult` and `Juvenile`. This was pretty useful, since

---

[2] http://insightsfromdata.io

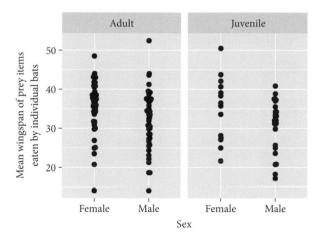

Figure 4.11 Difference in prey size in the diets of male/female adult/ juvenile bats.

now we have the words on the graph, rather than their abbreviations. Generally speaking, it's a good idea to use the full words in the data rather than codes, and here is one reason why. The graphs are easier to understand, need less explanatory text, and need less 'beautification'. This is an example of good preparation making things much, much easier quite some distance down the workflow.

Maybe you're wondering how we added the mean values to the graph in Figure 4.10? The first step is to calculate the means:

```
mean_prey_stats <- prey_stats %>%
  na.omit() %>%
  group_by(Sex, Age) %>%
  summarise(sex_mean_wingspan = mean(mean_wingspan, na.rm = TRUE))
```

Then we use the same code as before but add a `geom_point` layer using the dataset `prey_stats` that we recently made, mapping the *x*-axis to the `gm_mean_wingspan` variable within that dataset, and setting the size, colour, and transparency (`alpha`) of the points on the graph. (Our use of ggplot is getting quite complex… recall that all will be explained in the ggplot deeper dive in Chapter 7.)

```
prey_stats %>%
ggplot() +
  geom_beeswarm(mapping = aes(x = Sex, y = mean_wingspan)) +
  facet_wrap(~ Age) +
  geom_point(data = mean_prey_stats,
             mapping = aes(y = sex_mean_wingspan),
             size = 4,
             col = "blue",
             alpha = 0.5) +
  ylab("Mean wingspan of prey")
```

You can start to see how we can build quite nice graphs by adding one layer after another, and how we can use different datasets in different layers.

i What if we don't want the grey gridded background or the box around the graph? One way to deal with this is to change the 'theme' by adding + theme_classic(), for example, to our ggplot. Have a go at this. Have a look on the internet for other themes.

i Note that, above, we calculated the mean of the wingspans of the prey consumed by female bats by taking the mean of the mean wingspan of prey consumed by each bat. We did the same for the mean of the male bats. This is really important, as it implies we have properly accounted for non-independence in the data (the multiple prey found in one poop are non-independent). More about this is on the companion website.[3]

Beautiful! We can stop here. We've made some excellent insights about our data and drawn some gorgeous graphs.

---

[3] http://insightsfromdata.io

4.5 Another view of the question and data

4.5.1 BEFORE YOU CONTINUE...

Do you feel that you understand the material in the previous sections of the Workflow Demonstration well enough to continue? Please make sure you do. If you don't, then go back and look again; perhaps get together at a nice coffee shop/pub/bar/park with friends and take some relaxed time to discuss things. Make learning a pleasure; do it in a nice environment with people you like (Andrew and Natalie's preference), or do it alone in your special hiding place (Owen and Dylan's preference). And, really, feel free to email us if you need some help.

It is really important to feel that you understand what came before, because it will make what comes next much easier to deal with. We do give reminders about key things. But the risk of your carrying on regardless of not feeling you understand is that you then continue to feel like you understand less and less well, lose confidence, and fail to fall completely in love with R.

You may even decide to skip this part of the Workflow Demonstration... this will be fine... and continue now to the next chapter to solidify what you learned in the first part of the demonstration. Great! Then come back here.

4.5.2 A PREY-CENTRIC VIEW

The same data can often be looked at in multiple valid ways. So far, we have taken a 'bat-centric' perspective. We looked at the average wingspan of prey consumed by each bat, and also the proportion of migratory prey, and the number of prey. Each data point was a bat.

A different way to look at these data is to focus on individual prey species, and then to find the proportion of the male bat poops containing this prey. And the same for the female bats. And for each of the prey species. Take a peek at Figure 4.13 to see what we're aiming for.

In essence, we are recognizing and working with the binary nature of the data: that *each prey species was either found or not found in the poop of a particular bat.*

Let's just be totally clear about what we're aiming for by means of an example. We want to calculate that the prey species *Mythimna vitellina* (for example) was found in 17 out of the 69 female bat poops, and in 21 of the 74 male bat poops. And to get those numbers for each of the 115 prey species.

To get this prey-centric view, we must make some reasonably involved transformations of the data…

Transform the data

We need to transform and rearrange the dataset. We need a dataset where each combination of prey species and bat sex is a row, and which includes the column 'proportion of poops found in' (though we will not name it that!).

To get there takes a few steps. First, let's find the total number of prey species we're dealing with:

```
bats %>%
  summarise(num_prey = n_distinct(Species))
```

```
## # A tibble: 1 x 1
##    num_prey
##       <int>
## 1       115
```

So, there were a total of 115 different prey species found in the bat poops.

Next, we get the number of times each of the prey species was found in a female bat poop and a male bat poop:

```
num_poops <- bats %>%
  group_by(Sex, Species) %>%
  summarise(num_poops = n())
```

```
## `summarise()` regrouping output by 'Sex' (override with
## `.groups` argument)
```

This `group_by` and `summarise` gives us the number of rows in the dataset for each combination of `Species` (prey species) and the sex of the bat. We can see there are 165 rows in this new dataset. Is this what you were expecting? Did you have an expectation? We should have made a guess. We might have guessed there would be 230 rows, one for each of the 115 prey species, occurring twice, once for male and once for female bats. Can you figure out why there are 165 and not 230 rows?

The reason is that the dataset we imported only contains records of when a prey species was found in a poop, and does not contain absences. So our `num_poops` dataset is missing 65 prey species that were never found in a male poop, or never found in a female poop.

Let's check this out a bit more closely, finding out which prey were found only in female poops, which were found only in male poops, which were found in poops of both sexes, which were not found in males, and which were not found in females. Read the comments in the code to follow what we do. Here we will use what are called 'set operations' to do the work. The functions are in the **dplyr** package, so make sure this is loaded.

```
# Get list of all prey
all_prey <- num_poops %>%
  # force dplyr to ignore the grouping variable Sex in num_poops
  # so that we can get just the Species names
  ungroup() %>%
  select(Species) # 165 values ( = prey species)

# Get the prey species found in male poops:
prey_in_male_poops <- num_poops %>%
  filter(Sex == "Male") %>%
  # force dplyr to ignore the grouping variable Sex in num_poops
  # so that we can get just the Species names
  ungroup() %>%
  select(Species) # 93 values ( = prey species)

# and in female poops
prey_in_female_poops <- num_poops %>%
  filter(Sex == "Female") %>%
  # force dplyr to ignore the grouping variable Sex in num_poops
  # so that we can get just the Species names
  ungroup() %>%
  select(Species) # 72 values ( = prey species)
```

```
# Get the number of prey species found
# in either or both males and females...
# should be same as number of unique (i.e. 115 values)
prey_in_either_or_both <- union(prey_in_male_poops,
                                prey_in_female_poops) # 115 values

# Get the prey found in both males and females
prey_in_both <- intersect(prey_in_male_poops,
                          prey_in_female_poops) # 50 values

# Get the prey not found in male poops
prey_not_in_male <- setdiff(all_prey, prey_in_male_poops)
# 22 values

# and not found in female poops.
prey_not_in_female <- setdiff(all_prey, prey_in_female_poops)
# 43 values
```

So, we see that 50 prey were found in both, 22 were found in females but not males, and 43 in males but not females, giving 72 found in females, and 93 in males. These numbers all add up nicely. We have 165 rows in num_poops and are missing 22 + 43 = 65.

Let's now 'fix' the num_poops dataset so it contains rows for when a prey species was not found in a male or female poop, so that there are 230 rows. We are adding in the prey_not_in_female and prey_not_in_male species as 0's. Note that, based on our work above, prey_not_in_female and prey_not_in_male are already tibbles/data frames and have the species as a variable name:

```
num_poops <- bind_rows(num_poops,
                  tibble(Sex = "Female",
                  # this object is already a tibble with a
                          variable called Species
                          prey_not_in_female,
                          num_poops = 0),
                  tibble(Sex = "Male",
                  # this object is already a tibble with a
                          variable called Species
                          prey_not_in_male,
                          num_poops = 0)
)
```

We have used the function `bind_rows` to stick together three tibbles: the original `num_poops` dataset, a dataset of the prey not found in female bat poops, and a dataset of the prey not found in male bat poops. The result is a dataset with 230 rows—excellent. (Perhaps you should recap the section where we previously made a tibble, Section 3.6.1.)

In order to calculate the proportion of poops prey were found in, we need to add to this dataset the total number of male and of female poops prey were found in (here comes a lovely, though perhaps daunting, pipeline!):

```
total_num_poops <- bats %>%
  select(Sex, Bat_ID) %>%
  distinct() %>%
  group_by(Sex) %>%
  summarise(num_bats = n())
```

```
## `summarise()` ungrouping output (override with `.groups`
argument)
```

In words, we `select` only the two required variables (`Sex` and `Bat_ID`) to get only the `distinct` rows, and then `group_by` the `Sex` variable and count the number of bats (poops) of each sex with `summarise` and `n()`.

Look at the object `total_num_poops`. There are 69 poops from female bats, and 74 from male ones.

We now have the dataset `num_poops` containing the number of poops each prey was found in. And we have the totals, but in a different dataset. We need to combine these datasets together. That is, we need to join (merge) the two datasets. When we do this, we need to ensure the two datasets share at least one variable in common (they do: `Sex`). We use the function `full_join`:

```
bat_props <- full_join(num_poops, total_num_poops)
```

```
## Joining, by = "Sex"
```

This `full_join` function says join (or merge) the two datasets, keeping records that occur in one or the other or both datasets (other options are `left_join`, `right_join`, and `inner_join`; see the information box below for details). We get a message back `Joining, by = "Sex"` and get a new dataset `freqs` with 230 rows as before, but with a new column/variable containing 69 for females and 74 for males. How did `full_join` know to use the `Sex` variable to do the join? It assumed this was what we wanted it to do, and made this assumption because both the `num_poops` and the `total_num_poops` datasets have a variable called `Sex`. So `full_join` (and the other three joins) looks for variables in common and then does the merge by them. We can, if we like, tell it which variables to merge by, for example if they have different names in the two datasets.

The four joins. Merging together two datasets is a quite useful and common task. We will do it again in other Workflow Demonstrations; you will probably do it yourself quite a bit. There are four types of join/merge in **dplyr**. We just used `full_join`, which keeps all records from each dataset, even if there is no match; if there is no match, we get some NAs in the new row. `inner_join` keeps only the records for which there is a match. `left_join` keeps all the records/observations in the first dataset, regardless of match. `right_join` keeps all the observations in the second dataset, regardless of match. There are two other joins, `semi_join` and `anti_join`, but we won't cover them in this book.

When joining/merging datasets, always try to figure out what you expect before you do it. How many rows do you expect in the merged dataset, and why? R will do its best to join them, but is it doing what you want and what you think it should be doing? Make sure you know. Don't trust R to figure such important things out for you.

Awesome! We have a dataset (`bat_props`) with a row for each prey species × bat sex combination (230 rows), and columns including the number of poops the prey was found in and the total number of poops. We can get the proportion of poops each prey was found in with a `mutate`:

```
bat_props <- mutate(bat_props, props = num_poops / num_bats)
```

Looking at this dataset we can see, for example, that the prey species most commonly found in poops was *Autographa gamma/pulchrina*, appearing in 30 of the 69 (44%) female poops. And there are 65 prey species that were never found in some male or some female poops (good, as this is what we previously saw).

Great! We now have the dataset as we need it for further exploration.

Visualizing the proportions

Let's make a graph showing the proportion of poops each prey species was found in, separately for male and female bats. The following code produces Figure 4.12:

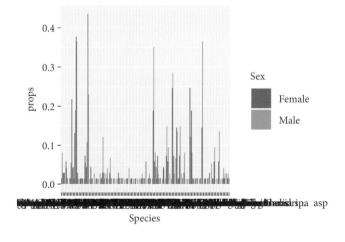

Figure 4.12 Frequencies with which prey species appeared in the diets of bats, way too small to see anything useful.

```
bat_props %>%
  ggplot() +
    geom_col(mapping = aes(x = Species, y = props, fill = Sex),
             position = "dodge")
```

This is a mess; there are just too many prey species to see. There are 230 bars, and all the prey species names are overlapping. Let's focus on the 10 prey species with the greatest difference in proportion between male and female bat poops. To find these 10 species we're going to use a pipeline of five **dplyr** functions: select %>% pivot_wider %>% mutate %>% arrange %>% top_n. The ones we haven't so far dealt with are pivot_wider (see more below) and top_n.

We use pivot_wider here to put the proportions for male poops and female poops into two separate columns, so we can then easily subtract one from another using mutate. You should be able to figure out the other steps. But just in case, we explain below the code:

```
topten_species <- bat_props %>%
  select(Species, Sex, props) %>%
  pivot_wider(names_from = Sex, values_from = props) %>%
  mutate(diff = abs(Female - Male)) %>%
  arrange(desc(diff)) %>%
  top_n(10)
```

```
## Selecting by diff
```

- select only the columns we need to work with; this helps with pivot_wider. We only need Species, Sex, and props.
- pivot_wider takes the entries in one variable and spreads them out into multiple columns. The pivot_wider function is useful when we need our data in wide format (like here). Look at Figure 6.2 to visually see what pivot_wider does. When we use it, we give it the variable that contains the groups that will be the different columns in the new dataset (Sex in the example above), and the variable that contains the information to be spread (odds in the example above).

- `mutate` to create a new variable `diff`, containing the absolute value of the difference between the proportions in male and female bat poops.
- `arrange` the dataset into descending order of the `diff` variable.
- use `top_n(10)` to grab the top 10, noting how the last function delivers the note that it is selecting by `diff`, our ranking of interest!

Here are the 10 species found:

```
topten_species
```

```
## # A tibble: 10 x 4
##    Species                          Female   Male  diff
##    <chr>                             <dbl>  <dbl> <dbl>
## 1 Rhodometra sacraria               0.145 0.365  0.220
## 2 Autographa gamma/ pulchrina       0.435 0.230  0.205
## 3 Agrotis ipsilon                   0.217 0.0405 0.177
## 4 Hoplodrina ambigua/ superstes     0.188 0.351  0.163
## 5 Peridroma saucia                  0.246 0.122  0.125
## # ... with 5 more rows
```

Now we can replot the previous figure but only for these 10 species with the greatest difference between the frequency of appearance in male and female poops. A key step is **subsetting** the `bat_props` data to get only the species listed in `topten_species`. We introduce another `_join` function (these are soooo cool) that does this: `semi_join`.

`semi_join` returns all rows from *x* with a match in *y*. You specify the variable by which to join.

The following code uses `x = bat_props`, `y = topten_species`, and `by = "Species"` to get the subset and produce Figure 4.13 by first

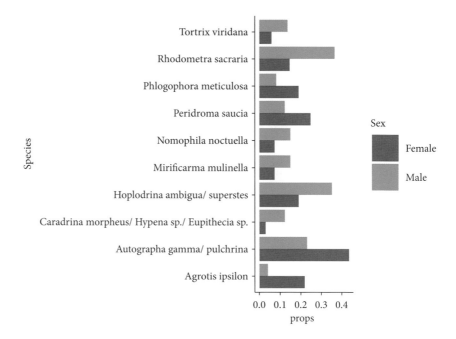

Figure 4.13 Frequencies with which prey species appeared in the diets of bats, only for the 10 species with the greatest difference between male and female bats.

using the `filter` function to keep only observations for the top 10 species (the filter function is covered in detail in Section 5.1.4):

```
semi_join(bat_props, topten_species, by = "Species") %>%
  ggplot() +
  geom_col(mapping = aes(x = Species, y = props, fill = Sex),
           position = "dodge") +
  coord_flip() +
  theme_classic(base_size = 7)
```

That's better. We can clearly see some prey species appear much more frequently in male bat poops, and some more frequently in female poops. This strongly suggests that male and female bats differ in their diets. Though we mustn't forget that here we selected only the 10 prey species that showed the greatest difference between male and female bats.

Odds and odds ratios

There were 69 poop samples from female bats, and 74 from males. For each prey species, a poop could contain that prey species or not contain it. For example, the prey species *Tortrix viridana* was found in 4 female poops and not found in 65 female poops. Hence the probability of finding this species in a female poop was 4/69 = 0.06. The probability of finding it in a male poop was 10/74 = 0.14.

Another way to compare the male and female diets is to compare the odds. Odds are the number of one outcome divided by the number of other outcomes. So the odds of finding *Tortrix viridana* in a female poop are 4/65 = 0.06. And for a male poop, the odds are 10/64 = 0.16. To compare these two numbers we typically divide one by the other, to obtain an *odds ratio*—here 0.06/0.16 = 0.38 (or 2.78 if we do the division the other way round). This means that the odds of finding this prey in a male poop are nearly three times as great as of finding it in a female poop.

Let's use R to get the odds ratio for each of the prey species. First we calculate the odds within a mutate, then we select only the variables we need, and then we pivot_wider to spread these data so that the odds for males are in one column and those for females in another. This will allow us to easily compute the odds ratio. We know there will be some divisions by zero, and therefore some funky results:

```
odds_ratios <- bat_props %>%
  mutate(not = num_bats - num_poops,
         odds = num_poops / not) %>%
  select(Sex, Species, odds) %>%
  pivot_wider(names_from = Sex, values_from = odds) %>%
  mutate(Odds_ratio = Female / Male)
```

Here is a glimpse of the odds ratios. The odds ratio for *Acrobasis bithynella* is 0.17. That is, it was much less likely to be found in the poop of a female bat than in the poop of a male bat.

```
glimpse(odds_ratios)
```

```
## Rows: 115
## Columns: 4
## $ Species    <chr> "Acrobasis bithynella", "Acrobasis obliqua", "Agonopteri...
## $ Female     <dbl> 0.01470588, 0.02985075, 0.02985075, 0.06153846, 0.014705...
## $ Male       <dbl> 0.08823529, 0.02777778, 0.00000000, 0.04225352, 0.000000...
## $ Odds_ratio <dbl> 0.1666667, 1.0746269, Inf, 1.4564103, Inf, 0.2573529, 6....
```

Now we can plot the odds ratio for the top 10 prey species with the greatest difference between the male and female poops, again using our lovely `semi_join`. Actually, let's plot the log to base 2 (`log2()`) of the odds ratio (we explain why we use $\log_2$ below). The following code produces Figure 4.14:

```
semi_join(odds_ratios, topten_species, by = "Species") %>%
  ggplot() +
  geom_point(mapping = aes(x = Species, y = log2(Odds_ratio))) +
  geom_hline(yintercept = 0, linetype = "dashed") +
  coord_flip() +
  theme_classic(base_size = 7)
```

Why do we care about odds ratios? Wikipedia (at least at one time) stated 'The odds ratio (OR) is a statistic defined as the ratio of the odds of A in the presence of B and the odds of A without the presence of B. This statistic attempts to quantify the strength of the association between A and B.' In our example, 'A in the presence of B' corresponds to a prey species being found (A) in a poop of a female bat (B), whereas 'A without the presence of B' corresponds to prey being found (A) in the absence of B (in the poop of a male bat). An odds ratio of 1 shows no difference in A (prey present) between the presence or absence of B (female or male). Another way to think about this is that the odds ratio measures whether the occurrence of A (presence of a prey species in a poop) is associated with the occurrence of B (whether the poop is from a female bat).

So why did we use $\log_2$ (odds ratio)? Imagine we play a game 10 times. We win 4 and lose 6. The odds of winning are 4/6. However, the odds against winning range from 0 to 4 (the worse we play, the closer we get to 0), while the odds in favour of winning range from 4 to infinity (the better we play, the further away from 4 we get). The odds against winning

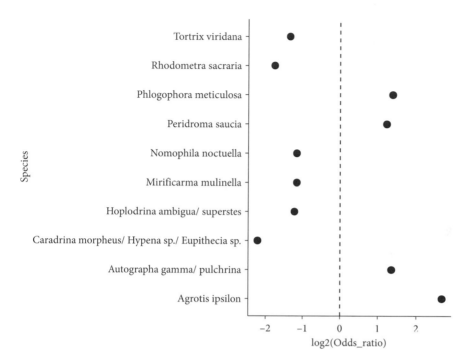

Figure 4.14 Odds ratio of frequencies for male and female bats consuming 11 prey species. Odds ratios greater than zero indicate that females are more likely to consume the prey species than male bats (and odds ratios smaller than zero indicate that males are more likely to consume the prey species than females).

are quite constrained relative to the odds in favour—there is asymmetry. The $\log_2(\text{odds})$ manages this asymmetry for us.

The odds ratio is quite central to data of the type we've been working with: binary data. If you learn about statistical models and inference for binary data, you will experience again and again the odds ratio.

4.6 A caveat

This Workflow Demonstration and those on the *Insights* companion website probably give the impression that one can think through all the likely issues with the data and, in a lovely linear process, work from the data to the insights. For example, we do things early on in the data manipulation in order to make things easier later on. It looks like we anticipate the problems that will arise and prevent them from ever occurring.

The truth is, however, that we often find problems once we are quite far into the data processing, and then realize we should go back and fix their causes (rather than fix their symptoms). An example is when we found the problem with NAs not being imported properly, and then went back to refine the import.

In another Workflow Demonstration we find a problem while plotting the data (the order of levels in the factor variable `composition`), realize this is caused by an issue that could be sorted out much earlier on, and therefore we fix it then. This has positive effects on all subsequent analyses.

In both cases, we did not treat the symptom... we fixed the source of the problem. This is a very, very good habit to get into. Fix the cause; don't just treat the symptom. Fixing the cause will increase efficiency, reliability, and all-round happiness. (Another example of our workflow appearing linear, but originally moving forwards, then back to improve or correct something, and then moving forwards again, is in the food-diversity-polity Workflow Demonstration on the *Insights* companion website[4]).

4.7 Summing up and looking forward

We covered a lot of ground here. And we finally got our insights. We discovered, for example, that female bats eat larger prey on average than males. We made some graphs to clearly show this. We also looked at the data from a different perspective and showed that the composition of prey species taken differs between female and male bats.

We also experienced a lot of new R and many new concepts. Some of the new R was *very* cool (at least in our opinion): `group_by` and `summarise` are brilliant, and `ggplot` is the best! But, as we said at the beginning of this chapter, don't worry too much if some of the R and some of the concepts

---

[4] http://insightsfromdata.io

did not fully sink in, or even if they bounced straight off you. In the next chapters we will look at these in more detail.

Recall that the *Insights* companion website[5] provides all the R scripts from this chapter and the previous one in one place, without any explanation or narrative.

First, though, we suggest that you again review and consolidate what you just experienced. Make a list of the new functions we used, what they do, their arguments, and the package they belong to. Make a list of the sections and, for each, comment on how much you understood. Go back to sections that were unclear to you, and look for or ask for help in understanding them better.

4.8 A small reward, if you like dogs

Try running this:

```
install.packages("devtools")
devtools::install_github("melissanjohnson/pupR")
library(pupR)
pupR()
```

We might give you a cat-related treat later!

---

[5] http://insightsfromdata.io

Dealing with data 1
*Digging into **dplyr***

In the previous two chapters we demonstrated a data analysis workflow to understand variation in the diets of bats. We saw a lot of R, and a lot of concepts, but we didn't go into huge detail about these. The aim was for to get through the whole workflow while having a reasonable chance of keeping your head above water. In the next few chapters we will take a deeper (but not too deep) dive into the details of R. You can perform the basic workflow without this information, but we highly recommend trying to understand it, as it will help you when working with your own data.

In this chapter you will become much better acquainted with the wonderful world of the **dplyr** package. We will look more carefully at the some of the core **dplyr** functions: `select` (get some columns), `mutate` (make a new column), `filter` (get some rows), `arrange` (order the rows), `group_by` (add grouping information), and `summarise` (calculate summary information about groups). Again, we advise you to make a list of these functions, and others you experience, and make notes about them.

Insights from data with R: An Introduction for the Life and Environmental Sciences. Owen L. Petchey, Andrew P. Beckerman, Natalie Cooper and Dylan Z. Childs, Oxford University Press (2021). © Owen L. Petchey, Andrew P. Beckerman, Natalie Cooper and Dylan Z. Childs. DOI: 10.1093/oso/9780198849810.003.0005

We also remind you to be kind to yourself! Make a plan of how you will tackle this chapter. For example, work for 45 minutes, take a 15 minute walk outside, work for 45 minutes, do some breathing exercises, work for 45 minutes, have a snack. And so on, until even after the 15 minute break your brain still feels like a big mess of R—at which point stop for the day, and do something else. (But before you stop, make a note of where you were and what you were trying to do, so that you can more easily pick up from that point.)

5.1 Introducing dplyr

If you have previous experience of R, you may have seen and/or used dollar signs and square brackets ($, [[, and [) to work with datasets in R. There are also a number of base R functions that allow us to manipulate datasets in a slightly more intuitive way. For example, there is a function called `transform` that adds new variables and changes existing ones, and a function called `subset` to select variables and subset rows in a data frame according to the values of its variables.

You did not see us use any of these in the Workflow Demonstration, and you won't see them elsewhere in this book. Instead, we rely on the **tidyr** and **dplyr** packages and their functions to get parts of a dataset, to transform variables, to create summaries, and to do many other useful things. The **dplyr** package provides a more elegant, efficient, readable, less error-prone approach to manipulating data. It is carefully designed to make life easy! It is *very* consistent in the way its functions work—the first argument of the main **dplyr** functions is always a dataset. Consistency makes for reliability, efficiency, readability, and beauty.

Another nice feature of **dplyr** is that each of the core functions does one thing well. These functions are often referred to as 'verbs'—they 'do something' to data. In the Workflow Demonstration you saw `select` (obtain a subset of variables, or exclude some variables), `mutate` (add new variables), `group_by` (state which variable the data should be grouped by), and `summarise` (calculate information about groups of rows). Another function, `filter`, gives a subset of rows, and another, `arrange`, reorders rows.

As well as being pretty elegant, **dplyr** is fast, and so a good option for quite large datasets. It also plays well with a number of database systems. Bottom line: learning **dplyr** gives you tools that continue to work even if you get much more advanced in your data analyses.

The dplyr cheat sheet[1] that RStudio (the company) has produced is very useful, summarizing the main tools in **dplyr**. Download it, print it, and keep a copy by your computer.

Let's go into a bit more detail about some **dplyr** functions.

5.1.1 SELECTING VARIABLES WITH THE `select` FUNCTION

Start by switching to your bat diet project, open the script file, run all the code, move to the bottom of the script, and make a new comment about what we're about to do, e.g. 'Here comes some practice with key dplyr processes'.

We use `select` to *select variables* from a data frame or tibble, for example when we need to work with only a few of the variables in a dataset. Basic usage of `select` looks like this:

```
data_set %>%
  select(variable_name1, variable_name2, ...)
```

Take note: this is not an example we can run. This is a 'pseudocode' example, designed to show, in abstract terms, how we use `select`:

- The data (`data_set`) are piped into the first argument position of the `select` function.
- We then include a series of one or more additional arguments, where each one is the name of a variable in `data_set`.

We've expressed the additional arguments as `variable_name1`, `variable_name2,...`, where `variable_name1` and `variable _name2` are names of two variables in `data_set` and`...` is a placeholder for any other variables.

For example, in the Workflow Demonstration we did

```
bats %>%
  select(Wingspan_mm)
```

to look at just the `Wingspan_mm` variable.

A few things about `select`:

- We do not have to surround variable names with quotes. The only time we need to break this convention is when a variable name has spaces in it (which is best avoided if you want to keep life simple).
- It does not have 'side effects', which is just a fancy way of saying it does not change the original `bats` data. Because we did not assign the result of `select` a name with the assignment arrow (`<-`), the result was printed to the Console—nothing else happened.
- The order of variables in the arguments determines the order in the new dataset (i.e. the column order). This means we can reorder variables at the same time as selecting them. Note that we are not talking here about sorting the rows of the data; this is done with the `arrange` function (Section 5.1.5).
- The `select` function returns the same kind of thing that we fed into it—a tibble if we gave it a tibble or a data frame if we gave it one of those.
- If we want to keep all variables but one, it is more convenient to specify the one we do *not* need. Use the minus operator (−) to indicate that that variable should be dropped. We did this in the Workflow Demonstration:

```
bats <- bats %>%
  select(-Date)
```

We just 'overwrote' the `bats` dataset with a new version that does not contain the `Date` variable. Be careful when you overwrite a dataset. And be aware the code just used to overwrite it will likely not work again unless all the preceding code is again run (e.g. because there is now no `Date` variable to remove).

The `select` function can also be used to keep a set of variables that occur consecutively in the dataset. If we wanted to keep the variables in the bat diet dataset from columns `Order` to `Family`, we would do

```
bats %>%
  select(Order:Family)
```

There are a number of helper functions for use within the `select` function, e.g. `starts_with`, `ends_with`, `contains`, `matches`. So, if we want to keep all the variables that start with 'S' or 's', we do the following:

```
bats %>%
  select(starts_with("s"))
```

```
## # A tibble: 633 x 3
##    Sex     Sp._Nr. Species
##    <chr>     <dbl> <chr>
## 1 Female       52 Ethmia bipunctella
## 2 Female       22 Bradycellus verbasci
## 3 Female       64 Hoplodrina ambigua/ superstes
## 4 Female       98 Rhyacia simulans
## 5 Female      114 Xestia c-nigrum
## # ... with 628 more rows
```

There is a nice help file you can review by typing `?starts_with` into the Console—choose the help file associated with *tidyselect*! Note that these **dplyr** helper functions are, by default, case *in*sensitive (unlike most other situations in R). If you'd like case sensitivity, add the argument `ignore.case = FALSE`:

```
bats %>%
  select(starts_with("s", ignore.case = FALSE))
```

```
## # A tibble: 633 x 0
```

5.1.2 RENAMING VARIABLES WITH select AND rename

The select function can also rename variables it keeps. Oops! We made a big deal above that **dplyr** functions do only one thing! select is a bit of an oddball. We can do this like so, with the new name on the left of the = and the old one on the right:

```
bats %>%
  select(Order, Sp_Nr = Sp._Nr.)
```

```
## # A tibble: 633 x 2
##    Order        Sp_Nr
##    <chr>        <dbl>
## 1 Lepidoptera     52
## 2 Coleoptera      22
## 3 Lepidoptera     64
## 4 Lepidoptera     98
## 5 Lepidoptera    114
## # ... with 628 more rows
```

If we want to keep all the variables and just rename a few, we use rename:

```
bats %>%
  rename(Sp_Nr = Sp._Nr.)
```

As in the select example, the convention is the new name on the left of the = and the old one on the right.

5.1.3 CREATING NEW VARIABLES WITH THE mutate FUNCTION

Adding or creating new variables is a common task in data visualization and analysis. We might want to create a log-transformation of a variable, or create a version of a date variable that can be used for analyses. In the Workflow Demonstration, we used the `mutate` function to add a new date variable:

```
bats <- bats %>%
  mutate(Date_proper = dmy(Date))
```

As usual with **dplyr** functions, the first argument is the dataset we want it to do its magic on. Subsequent arguments are the work to do. Here the second argument says 'please make a new variable called `Date_proper` and set it equal, =, to the result of `dmy(Date)`. Thanks!' (though **dplyr** doesn't actually care whether or not you thank it). It is worth noting again that there are no 'side effects' to the use of mutate—your input data are not affected; the new variable is only created in the copy made by the **dplyr** function.

> The **dplyr** package also contains the `transmute` function, which can be used to create new variables just like above, but then only keep the newly created variables.

Sometimes we want to change the values being used to define observations in a particular variable. For example, we might want to change abbreviations like 'M' and 'F' to 'Male' and 'Female' so we don't get confused. We used `mutate` and its friend the `case_when` function to substitute the values of the sex and age variables:

```
bats <- bats %>%
  mutate(Sex = case_when(Sex == "M" ~ "Male",
                         Sex == "F" ~ "Female"),
         Age = case_when(Age == "Ad" ~ "Adult",
                         Age == "Juv" ~ "Juvenile"))
```

The `case_when` helper is really valuable. You might have used nested-if-else statements in Excel to do these kinds of substitutions… and cried. This is a very clean, literate, and reproducible way to do these substitutions.

A few things to know about the `mutate` function…

- Here we have assigned the result of `mutate` back to the object `bats` ('overwriting' the previous copy of `bats`).
- Just like the `select` function, `mutate` returns a data frame if we give it a data frame, or a tibble if we give it a tibble.
- Within the same `mutate`, we can use newly created variables in further calculations. This is really nice, and quite special.

Notice that we put each argument on a new line, with a comma at the end of each line to separate the arguments. R ignores white space like these new lines, but they are useful because long lines of code can be broken onto new lines to make everything more human-readable. Really try to do this, especially so you don't have to scroll right to see the end of a long line of code. You will love yourself for keeping your code lines short and within the width of your script editor.

Pay close attention here to the indentation. Each of the four arguments starts at the same position horizontally. RStudio did this for us after we pressed the Return key at the end of the line. This **automatic indentation** is a very nice feature. It can help us spot when something isn't quite right, because then the indentation will not be quite right. For example, if we type this code but forget a bracket after `Date` we get

```
bats <- bats %>%
  mutate(Date_proper = dmy(Date,
                           Sex = case_when(Sex == "M" ~ "Male",
                                           Sex == "F" ~ "Female"),
                           Age = case_when(Age == "Ad" ~ "Adult",
                                           Age == "Juv" ~ "Juvenile")))
```

The word `Sex` is now indented as if it is within the `dmy` function (because it is!), but we know we don't want that. Because we often

forget brackets, paying attention to indentation really pays off. *If the automatic indentation is wrong, chances are you forgot a bracket somewhere.*

If we edit code after we first write it then the indentation can get thrown off. We can have RStudio reapply its automatic indentation by selecting some code and pressing Cmd-I (Mac) or Ctrl-I (Windows). You can also go to the menu item Code > Reindent Lines.

5.1.4 GETTING PARTICULAR OBSERVATIONS WITH filter

We use filter to *subset observations* in a data frame or tibble. This is a really common activity in exploring data. We might want to see observations where Wingspan_mm is larger or smaller than some value, or we might want to see only data on females. This is often done when we want to explore only some of the observations in a dataset.

We used the filter function in the Workflow Demonstration when we wanted to plot only the top 10 prey species. Basic usage of filter is in this *pseudocode* (i.e. it's not an example we can run):

```
data_set %>%
  filter(condition1,
         condition2, ...)
```

We first pipe data_set into the filter function, as with most **dplyr** functions. Then come one or more additional arguments, where each of these is an R expression—a little snippet of R code—involving one or more variables in data_set. Each expression must give an answer of TRUE or FALSE for each value of the variable. We've expressed these as condition1, condition2, ..., where condition1 and condition2 represent the first two conditions, and the ... is acting as a placeholder for any other conditions.

That's pretty abstract! To see `filter` in action, we'll use it to get some observations from the bat diet dataset based on three expressions (criteria):

```
bats %>%
  filter(Age == "Adult",
         Sex == "Female",
         Wingspan_mm < 15)
```

```
## # A tibble: 7 x 15
##   Row_order Bat_ID Age   Sex    Sp._Nr. Species Class Order Family Pest
##       <dbl>  <dbl> <chr> <chr>    <dbl> <chr>   <chr> <chr> <chr>  <chr>
## 1       153   1193 Adult Fema~       72 Mirifi~ Inse~ Lepi~ Gelec~ no
## 2       173   1203 Adult Fema~       57 Eupith~ Inse~ Lepi~ Geome~ no
## 3       317   1539 Adult Fema~       72 Mirifi~ Inse~ Lepi~ Gelec~ no
## 4       328   1581 Adult Fema~       41 Crocid~ Inse~ Lepi~ Tortr~ no
## 5       484   1807 Adult Fema~       95 Pyraus~ Inse~ Lepi~ Cramb~ no
## # ... with 2 more rows, and 5 more variables: Migratory <chr>,
## #   Wingspan_mm <dbl>, No._Reads <dbl>, Date <chr>, Date_proper <date>
```

In this example we've created a subset of `bats` that only includes the seven observations where the `Age` variable is equal to `Adult` AND the `Sex` variable is equal to `Female` AND the `Wingspan_mm` variable is less than 15. All three conditions must be met for an observation to be included in the resulting tibble.

This is probably starting to become repetitious, but there are a few features of `filter` that we should note:

- As usual, the result produced by `filter` in our example was printed to the Console. The `filter` function did not change the original `bats` in any way (no side effects!).
- The `filter` function returns the same kind of data object as the one it is working on: it returns a data frame if our data were originally in a data frame, and a tibble if they were in a tibble.

We can achieve the same result as in the above example in a different way. This involves the & operator:

```
bats %>%
  filter(Age == "Adult" &
         Sex == "Female" &
         Wingspan_mm < 15)
```

Once again, we have created a subset of `bats` that only includes observations for adult female bats eating moths with wingspans less than 15 mm. This time, however, we used & to link the three criteria, so, again, observations are only retained if all three are true.

Sometimes we'd like to get observations on an either/or basis, for example if we want small and large moth wingspans. To do this, we use a different operator, one that specifies OR rather than AND. For example,

```
bats %>%
  filter(Wingspan_mm < 10 |
         Wingspan_mm > 50)
```

```
## # A tibble: 46 x 15
##   Row_order Bat_ID Age   Sex   Sp._Nr. Species Class Order Family Pest
##       <dbl>  <dbl> <chr> <chr>   <dbl> <chr>   <chr> <chr> <chr>  <chr>
## 1         4    367 Adult Fema~      98 Rhyaci~ Inse~ Lepi~ Noctu~ no
## 2        27    606 Adult Male       72 Mirifi~ Inse~ Lepi~ Gelec~ no
## 3        33    716 Juve~ Male       72 Mirifi~ Inse~ Lepi~ Gelec~ no
## 4        40    725 Juve~ Male       72 Mirifi~ Inse~ Lepi~ Gelec~ no
## 5        83    841 Adult Male       80 Noctua~ Inse~ Lepi~ Noctu~ no
## # ... with 41 more rows, and 5 more variables: Migratory <chr>,
## #   Wingspan_mm <dbl>, No._Reads <dbl>, Date <chr>, Date_proper <date>
```

This creates a subset of `bats` that only includes the 46 observations where bats ate moths with wingspans less than 10 OR greater than 50 mm. The vertical bar | is understood by R to mean OR.

We might also be looking for observations between a set of values. A helper function useful with `filter` is the `between` function. This is used to identify the values of a variable that lie inside a defined range:

```
bats %>%
  filter(between(Wingspan_mm, 15, 17))
```

```
## # A tibble: 5 x 15
##   Row_order Bat_ID Age   Sex   Sp._Nr. Species Class Order Family Pest
##       <dbl>  <dbl> <chr> <chr>   <dbl> <chr>   <chr> <chr> <chr>  <chr>
## 1       125   1132 Adult Male       23 Bryotr~ Inse~ Lepi~ Gelec~ no
## 2       156   1193 Adult Fema~      61 Gymnos~ Inse~ Lepi~ Geome~ yes
## 3       366   1710 Adult Male       23 Bryotr~ Inse~ Lepi~ Gelec~ no
## 4       373   1710 Adult Male       61 Gymnos~ Inse~ Lepi~ Geome~ yes
## 5       430   1744 Juve~ Male       61 Gymnos~ Inse~ Lepi~ Geome~ yes
## # ... with 5 more variables: Migratory <chr>, Wingspan_mm <dbl>,
## #   No._Reads <dbl>, Date <chr>, Date_proper <date>
```

This example filters the `bats` dataset such that only values of `Wingspan_mm` between and including 15 and 17 are retained. We could do the same thing with `Wingspan_mm >= 15 & Wingspan_mm <= 17`, but the `between` function makes things a bit easier to read. Furthermore, with a quick 'negation' `!`, we can ask for the observations where `Wingspan_mm` is not between 15 and 17:

```
bats %>%
  filter(!between(Wingspan_mm, 15, 17))
```

```
## # A tibble: 550 x 15
##   Row_order Bat_ID Age    Sex    Sp._Nr. Species Class Order Family Pest
##       <dbl>  <dbl> <chr>  <chr>    <dbl> <chr>   <chr> <chr> <chr>  <chr>
## 1         1    366 Adult  Fema~       52 Ethmia~ Inse~ Lepi~ Depre~ no
## 2         3    367 Adult  Fema~       64 Hoplod~ Inse~ Lepi~ Noctu~ no
## 3         4    367 Adult  Fema~       98 Rhyaci~ Inse~ Lepi~ Noctu~ no
## 4         5    367 Adult  Fema~      114 Xestia~ Inse~ Lepi~ Noctu~ yes
## 5         6    367 Adult  Fema~       19 Autogr~ Inse~ Lepi~ Noctu~ yes
## # ... with 545 more rows, and 5 more variables: Migratory <chr>,
## #   Wingspan_mm <dbl>, No._Reads <dbl>, Date <chr>, Date_proper <date>
```

Here's another useful operator and criterion (we used it to get the top 10 species). The group membership `%in%` operator (part of base R, not **dplyr**) is used to determine whether the values in one variable occur among the values in a second. It's used like this: `vec1 %in% vec2`. This returns a vector where the values are `TRUE` if an element of `vec1` is in `vec2`, and `FALSE` otherwise. We can use the `%in%` operator with `filter` to get observations/rows by the values of one or more variables:

```
bats%>%
  filter(Order
  %in% c("Hemiptera", "Coleoptera", "unclassified Insecta"))
```

```
## # A tibble: 12 x 15
##   Row_order Bat_ID Age    Sex    Sp._Nr. Species Class Order Family Pest
##       <dbl>  <dbl> <chr>  <chr>    <dbl> <chr>   <chr> <chr> <chr>  <chr>
## 1         2    366 Adult  Fema~       22 Bradyc~ Inse~ Cole~ Carab~ no
## 2        24    606 Adult  Male        84 Nysius~ Inse~ Hemi~ Lygae~ no
## 3        34    716 Juve~  Male        84 Nysius~ Inse~ Hemi~ Lygae~ no
## 4        41    725 Juve~  Male        84 Nysius~ Inse~ Hemi~ Lygae~ no
## 5        91    873 Adult  Fema~       84 Nysius~ Inse~ Hemi~ Lygae~ no
## # ... with 7 more rows, and 5 more variables: Migratory <chr>,
## #   Wingspan_mm <dbl>, No._Reads <dbl>, Date <chr>, Date_proper <date>
```

Here we get the 12 rows where the `Order` variable is equal to either 'Hemiptera', 'Coleoptera', or 'unclassified Insecta'.

5.1.5 ORDERING OBSERVATIONS WITH `arrange`

We use `arrange` to *reorder the rows* of our data. There are instances when we *need* to do this, but they are not so common; more often we just like to reorder rows so that we can understand our data a bit more easily when we view them. Basic usage of `arrange` looks like this:

```
data_set %>%
  arrange(vname1, vname2, ...)
```

Yes, this is pseudocode. As always, we first pipe `data_set` into the function. We then include a series of one or more additional arguments, where each of these should be the name of a variable in `data_set`: `vname1` and `vname2` are names of the first two ordering variables, and the `...` is acting as a placeholder for any remaining variables.

To see `arrange` in action, let's construct a new version of `bats` where the rows have been reordered by two numeric variables, first by `No._Reads` and then by `Wingspan_mm`:

```
bats_sorted <- bats %>%
  arrange(No._Reads, Wingspan_mm)

## and glimpse only a few relevant variables
bats_sorted %>%
  select(Dat_ID, No._Reads, Wingspan_mm) %>%
  glimpse()
```

```
## Rows: 633
## Columns: 3
## $ Bat_ID      <dbl> 1436, 1743, 1900, 1727, 606, 1887, 1741, 1829, 1321, 11...
## $ No._Reads   <dbl> 190, 250, 255, 269, 274, 300, 314, 332, 336, 338, 341, ...
## $ Wingspan_mm <dbl> 20.5, 29.5, 47.5, NA, NA, 29.5, 35.0, 31.0, 20.0, 6.5, ...
```

In `bats_ordered`, the rows are sorted first according to the values of `No._Reads` and then `Wingspan_mm`, each in ascending order, i.e. from smallest to largest. Since `No._Reads` appears before `Wingspan_mm` in the arguments, the values of `Wingspan_mm` are only used to break ties within any particular value of `No._Reads`.

For the sake of avoiding any doubt about how `arrange` works, let's quickly review its behaviour:

- The `arrange` function did not change the original `bats` data in any way.
- The `arrange` function returns the same kind of data object as the one it is working on.

There is one more thing to learn about `arrange`. By default, it sorts variables in ascending order. If we need it to sort a variable in descending order, we can wrap the variable name in the `desc` function:

```
bats_sorted <- bats %>%
  arrange(desc(No._Reads),
          desc(Wingspan_mm))
## and glimpse only a few relevant variables
bats_sorted %>%
  select(Bat_ID, No._Reads, Wingspan_mm) %>%
  glimpse()
```

```
## Rows: 633
## Columns: 3
## $ Bat_ID      <dbl> 1712, 1320, 1934, 1925, 1469, 890, 719, 1479, 1935, 732...
## $ No._Reads   <dbl> 135881, 49236, 45937, 45264, 44861, 43986, 43499, 43043...
## $ Wingspan_mm <dbl> 48.5, NA, 25.0, 25.0, 38.5, 37.5, 37.5, 37.5, 25....
```

This creates a new version of `bats` where the rows are sorted according to the values of `No._Reads` and `Wingspan_mm` in descending order. If you prefer, you can precede a variable name with a minus sign (`-`) instead of using the `desc` function (the effect will be identical):

```
bats_sorted <- bats %>%
  arrange(-No._Reads,
          -Wingspan_mm)
```

When you look at a dataset using the `View` function, i.e. in the spreadsheet-like view in R, you can order the rows by a particular column/variable by clicking on the name of the variable. This is a quite useful feature; for example, we can quickly see the observation with the largest value of a variable. When we do this ordering we *do not*

affect the order of the data in `bats`. The data are only reordered on the screen, not in the object. Use the `arrange` function to arrange the rows of a dataset.

5.2 Grouping and summarizing data with **dplyr**

We very often want to calculate things about groups of observations, such as their mean or median (we did so when we calculated the average wingspan of prey taken by female and by male bats). This is really common because we are often interested in comparing responses among groups (like prey size between bat sexes or between bat ages or between moth migratory status).

To see how this works in more detail, let us first consider the group of observations that represents our entire dataset. For example, we found the number of distinct bats in the entire bat diet dataset via

```
bats %>%
  summarise(n_distinct(Bat_ID))
```

Now consider when the groups are subsets of observations, as was the case when we found the number of bats in each sex and age class:

```
bats %>%
  group_by(Sex, Age) %>%
  summarise(n_distinct(Bat_ID))
```

You probably noticed that we used `group_by` and `summarise` quite a lot. They are very useful and very powerful. We use them a lot in our own work. They are a bit different from the **dplyr** functions we have already looked at, so it's worth taking a close look at them:

- The `group_by` function adds *grouping* information into a data object (e.g. a data frame or tibble), which makes subsequent calculations happen on a *group-specific* basis. It always spits out a tibble, because data frames aren't designed to hold grouping information.

- The `summarise` function is a data aggregation function that calculates summaries of one or more variables, separately for each group defined by `group_by`. Often these are single-value summaries, but the latest version of **dplyr** allows us to use multivalue summaries if we need to.

In the following sections, we dive a bit deeper, first into `summarise` and then into using it with `group_by`.

5.2.1 SUMMARIZING DATA—THE NITTY-GRITTY

We use `summarise` to *calculate summaries of variables* in an object containing our data. We do this kind of calculation all the time when analysing data. In terms of pseudocode, the usage of `summarise` looks like this:

```
data_set %>%
   summarise(expression1,
             expression2, ...)
```

Again, we first pipe `data_set` into the function. We then include a series of one or more additional arguments; each of these is a valid R expression performing an operation on at least one variable in `data_set`. These are given by the pseudocode placeholder `expression1`, `expression2`, ..., where `expression1` and `expression2` represent the first two expressions, and the ... is acting as a place holder for any other expressions.

These expressions can be any calculation involving R functions that return a vector of some kind. When it was first designed, `summarise` only worked with functions that return a single value. These days, that constraint no longer applies. The output is allowed to be a vector of any length. That said, most of the time, people use `summarise` with functions that spit out a single number (e.g. a mean).

With that in mind, and with the examples already given, you should be able to get the mean wingspan and number of reads in the entire bat diet dataset. Try to do this yourself before looking at the code immediately below (don't look!):

```
bats %>%
  summarise(mean_wingspan  = mean(Wingspan_mm, na.rm = TRUE),
            mean_num_reads = mean(No._Reads))
```

```
## # A tibble: 1 x 2
##   mean_wingspan mean_num_reads
##           <dbl>          <dbl>
## 1          33.6          7700.
```

Notice that we've provided informative names for ourselves on the left side of the =. Notice too what kind of object summarise returns here: it's a tibble with one row and two columns. There are two columns because we calculated two things. Simple, eh?!. There are a few other things to note about how summarise works:

- Even though the numbers of rows and columns have changed, summarise returns the same kind of data object as its input. It returns a data frame if our data were originally in a data frame, or a tibble if they were in a tibble.
- It's almost always better to name the new variables ourselves. If we don't, the names produced by summarise will be quite difficult to subsequently use. We do this by naming the arguments using =, placing the name we require on the left-hand side.
- The numbers printed to the Console are rounded, but they are only rounded when printed to the Console. If we assign the values of summarise to an object then the full-accuracy numbers are retained. This is really important, since often we want to do further calculations on the results of a summarise, and should do these on the full-accuracy numbers.

There are very many base R functions that can be used with summarise. Basically, anything that takes a vector as input and spits out another vector is probably OK. A few useful ones for calculating summaries of numeric variables are:

- `min` and `max` calculate the minimum and maximum values of a vector.
- `mean` and `median` calculate the mean and median of a numeric vector.
- `sd` and `var` calculate the standard deviation and variance of a numeric vector.

These all produce single-number summaries. What about multinumber summaries? As we said, these definitely work as well. Here is an example that uses `range` to calculate the minimum and maximum value of the wingspan (its 'range'):

```
bats %>%
  summarise(wingspan_range = range(Wingspan_mm, na.rm = TRUE))
```

```
## # A tibble: 2 x 1
##    wingspan_range
##             <dbl>
## 1             6.5
## 2            52.5
```

We can use several functions in `summarise`. The only 'rule' we have to follow is that they all return vectors of the same length. This means we can do arbitrarily complicated calculations in a single step. For example, we can calculate four things about the bats in the bat diet dataset:

```
bats %>%
  summarise(num_prey = n(),
            num_prey_distinct = n_distinct(Sp._Nr.),
            mean_wingspan = mean(Wingspan_mm, na.rm = TRUE),
            prop_migratory = sum(Migratory == "yes") / n())
```

Notice again that we placed each argument on a separate line in this example. This is just a style issue—we don't have to do this, but since R doesn't care about white space, we can use new lines and spaces to keep everything a bit more human-readable. It pays to organize `summarise` calculations like this as they become longer. It allows us to see the logic of the calculations more easily, and helps us spot potential errors when they occur.

Summarize all columns or types of columns. In the Workflow Demonstration (Section 3.7.8), we used the function `across` to count up the number of NAs in every column (by specifying `everything`). We'd like to explain it a bit more here, as it's a super-useful tool for generating summaries. Here is what we did:

```
bats %>%
  summarise(across(
    .cols = everything(),
    .fns = ~sum(is.na(.))))  %>%
  glimpse()
```

It is just quite a cool function and has two strangely expressed arguments, `.cols` and `.fns`, that allow you to specify columns in several ways and apply functions to these columns. The `.cols` argument is pretty clever—it is possible to have `across` identify, for example, all numeric variables, or any variable with the word `Bat` in it, and apply some function to these variables. Two examples are shown here, the first calculating the mean of all/any numeric variables, and the second finding the number of distinct values in columns with the word `Bat` in them:

```
# the mean of ALL numeric columns in the bats data
# where(is.numeric) hunts for numeric columns!
bats %>%
  summarise(
    across(.cols = where(is.numeric),
           .fns = ~mean(.))
  ) %>%
```

```
glimpse()

# the number of distinct Bat_IDs, as there is only 1 columns
# with the string "Bat"
bats %>%
  summarise(
    across(.cols = contains("Bat"),
           .fns = ~n_distinct(.))
  ) %>%
  glimpse()
```

5.2.2 GROUPED SUMMARIES USING group_by MAGIC

As noted above, performing a calculation with one or more variables over the whole dataset is useful, but very often we also need to carry out the same operations on different subsets of our data. For example, it's probably more useful to know how the mean wingspan varies among the different ages and sexes of bats in the bat diet dataset, rather than knowing the overall mean wingspan.

We could calculate separate means by using filter first to create different subsets of the data, and then using summarise on each of these subsets to calculate the relevant means. This would get the job done, but it's not very efficient and very soon becomes tiresome when we have to work with many groups.

The group_by function provides a more elegant, often even magical-seeming, solution to this kind of problem. It doesn't do all that much on its own, though. All the group_by function does is add a bit of grouping information—a meta-data tag—to a tibble or data frame. In effect, it defines subsets of data on the basis of one or more **grouping variables**. The magic happens when the grouped object is used with a **dplyr** verb like summarise or mutate. Once a data frame or tibble has been 'tagged' with grouping information, operations that involve these (and other) verbs are carried out separately on each of the defined groups of the data.

Basic usage of group_by looks like this:

```
data_set %>%
  group_by(vname1, vname2, ...)
```

Yet again (this is getting boring!), we first pipe `data_set` into the function. We then have to include one or more additional arguments, where each of these is the name of a variable in `data_set`. I have expressed this as `vname1, vname2, ...`, where `vname1` and `vname2` are names of the first two grouping variables, and the `...` is acting as a placeholder for any remaining variables.

As usual, it's much easier to understand how `group_by` works once we've seen it in action. You already saw it in action in the Workflow Demonstration, to calculate the number of prey items eaten by each bat:

```
prey_stats <- bats %>%
  group_by(Bat_ID) %>%
  summarise(num_prey=n())
```

```
## `summarise()` ungrouping output (override with `.groups` argument)
```

The first step is to use `group_by` to add grouping information to the dataset:

```
bats %>%
  group_by(Bat_ID) %>%
  glimpse()
```

```
## Rows: 633
## Columns: 15
## Groups: Bat_ID [143]
## $ Row_order   <dbl> 1, 2, 3, 4, 5, 6, 7, 8, 9, 10, 11, 12, 13, 14, 15, 16, ...
## $ Bat_ID      <dbl> 366, 366, 367, 367, 367, 367, 367, 367, 367, 367, 367, ...
## $ Age         <chr> "Adult", "Adult", "Adult", "Adult", "Adult", "Adult", "...
## $ Sex         <chr> "Female", "Female", "Female", "Female", "Female", "Fema...
## $ Sp._Nr.     <dbl> 52, 22, 64, 98, 114, 19, 2, 4, 70, 81, 29, 49, 19, 79, ...
## $ Species     <chr> "Ethmia bipunctella", "Bradycellus verbasci", "Hoplodri...
## $ Class       <chr> "Insecta", "Insecta", "Insecta", "Insecta", "Insecta", ...
## $ Order       <chr> "Lepidoptera", "Coleoptera", "Lepidoptera", "Lepidopter...
## $ Family      <chr> "Depressariidae", "Carabidae", "Noctuidae", "Noctuidae"...
## $ Pest        <chr> "no", "no", "no", "no", "yes", "yes", "no", "no", "yes"...
## $ Migratory   <chr> "no", "no", "no", "yes", "yes", "yes", "no", "no", "yes...
## $ Wingspan_mm <dbl> 23.5, NA, 31.0, 52.5, 40.0, 37.5, NA, 20.5, 39.0, 29.0,...
## $ No._Reads   <dbl> 32672, 411, 14796, 8953, 3472, 3282, 2178, 1807, 1067, ...
## $ Date        <chr> "20.07.12", "20.07.12", "20.07.12", "20.07.12", "20.07....
## $ Date_proper <date> 2012-07-20, 2012-07-20, 2012-07-20, 2012-07-20, 2012-0...
```

Compare this with the output produced when we print the ungrouped `bats` dataset:

```
bats %>%
  glimpse()
```

```
## Rows: 633
## Columns: 15
## $ Row_order   <dbl> 1, 2, 3, 4, 5, 6, 7, 8, 9, 10, 11, 12, 13, 14, 15, 16, ...
## $ Bat_ID      <dbl> 366, 366, 367, 367, 367, 367, 367, 367, 367, 367, 367, ...
## $ Age         <chr> "Adult", "Adult", "Adult", "Adult", "Adult", "Adult", "...
## $ Sex         <chr> "Female", "Female", "Female", "Female", "Female", "Fema...
## $ Sp._Nr.     <dbl> 52, 22, 64, 98, 114, 19, 2, 4, 70, 81, 29, 49, 19, 79, ...
## $ Species     <chr> "Ethmia bipunctella", "Bradycellus verbasci", "Hoplodri...
## $ Class       <chr> "Insecta", "Insecta", "Insecta", "Insecta", "Insecta", ...
## $ Order       <chr> "Lepidoptera", "Coleoptera", "Lepidoptera", "Lepidopter...
## $ Family      <chr> "Depressariidae", "Carabidae", "Noctuidae", "Noctuidae"...
## $ Pest        <chr> "no", "no", "no", "no", "yes", "yes", "no", "no", "yes"...
## $ Migratory   <chr> "no", "no", "no", "yes", "yes", "yes", "no", "no", "yes...
## $ Wingspan_mm <dbl> 23.5, NA, 31.0, 52.5, 40.0, 37.5, NA, 20.5, 39.0, 29.0,...
## $ No._Reads   <dbl> 32672, 411, 14796, 8953, 3472, 3282, 2178, 1807, 1067, ...
## $ Date        <chr> "20.07.12", "20.07.12", "20.07.12", "20.07.12", "20.07....
## $ Date_proper <date> 2012-07-20, 2012-07-20, 2012-07-20, 2012-07-20, 2012-0...
```

There is almost no change in the printed information—group_by really doesn't do much on its own. The main change is that the tibble resulting from the group_by operation has a little bit of additional information printed at the top: Groups: Bat_ID [143]. This tells us that the tibble is grouped by the Bat_ID variable. The [143] part tells us that there are 143 different groups. The only thing group_by did was add this grouping information to a copy of bats. If we actually want to do anything useful with the result, we need to assign it a name so that we can work with it.

Now we have a grouped tibble called bats_grouped, where the groups are defined by the values of Bat_ID. We see this in the second row of the output in R: # Groups: Bat_ID [143]. Any operations on this tibble will now be performed on a 'by group' basis. To see this in action, we use summarise to calculate the number of prey items eaten by each bat:

```
bats_grouped %>%
  summarise(num_prey = n())
```

```
## `summarise()` ungrouping output (override with `.groups` argument)

## # A tibble: 143 x 2
##     Bat_ID num_prey
##      <dbl>    <int>
## 1      366        2
## 2      367        9
```

```
## 3       584           6
## 4       598           5
## 5       606           6
## # ... with 138 more rows
```

```
# which is the same as
bats %>%
  group_by(Bat_ID) %>%
  summarise(num_prey=n())
```

```
## `summarise()` ungrouping output (override with `.groups` argument)
```

```
## # A tibble: 143 x 2
##    Bat_ID num_prey
##     <dbl>    <int>
## 1     366        2
## 2     367        9
## 3     584        6
## 4     598        5
## 5     606        6
## # ... with 138 more rows
```

To recap, when we used `summarise` on an ungrouped tibble the result was a tibble with one row: the overall global value. Now the resulting tibble has 143 rows, one for each value of `Bat_ID`. The `Bat_ID` variable in the new tibble tells us what these values are; the `num_prey` variable shows the number of prey of each individual bat.

Let's keep these calculated numbers for later use:

```
bats_num_prey <- bats_grouped %>%
  summarise(num_prey = n())
```

```
## `summarise()` ungrouping output (override with `.groups` argument)
```

5.2.3 MORE THAN ONE GROUPING VARIABLE

What if we need to calculate summaries using more than one grouping variable? The workflow is unchanged. Let's assume we want to know the mean number of prey eaten by male/female (the `Sex` variable) and juvenile/adult (the Age variable) bats. We first make a grouped copy of the dataset with the appropriate grouping variables. Notice that we are now working with the dataset we just calculated, which contains the number of prey eaten by each bat:

```
bats_num_prey_grouped <- bats_num_prey %>%
  group_by(Sex, Age)
```

```
## Error: Must group by variables found in `.data`.
## * Column `Sex` is not found.
## * Column `Age` is not found.
```

Hmmm... this gave the error `Column 'Sex' is not found`. This is because, when we did the previous `group_by-summarise`, we lost the `Sex` and `Age` variables. To keep this, we add them to the previous `group_by` and then proceed. This time, let's step back and simply link the three steps with three pipes (we term this a pipeline):

```
bats_num_prey_grouped <- bats %>%
  group_by(Bat_ID, Sex, Age) %>%
  summarise(num_prey = n()) %>%
  group_by(Sex, Age)
```

```
## `summarise()` regrouping output by 'Bat_ID', 'Sex' (override with `.groups` argument)
```

```
glimpse(bats_num_prey_grouped)
```

```
## Rows: 143
## Columns: 4
## Groups: Sex, Age [4]
## $ Bat_ID   <dbl> 366, 367, 584, 598, 606, 628, 716, 719, 725, 726, 727,
## $ Sex      <chr> "Female", "Female", "Female", "Male", "Male", "Female",
## $ Age      <chr> "Adult", "Adult", "Adult", "Adult", "Adult", "Adult",
## $ num_prey <int> 2, 9, 6, 5, 6, 3, 7, 1, 4, 5, 2, 1, 2, 7, 1, 6, 6, 11,
```

Good! We grouped the `bats_num_prey` data by `Sex` and `Age` and assigned the grouped tibble the name `bats_num_prey_grouped`. When we print this to the Console we see `Groups: Sex, Age [4]` near the top, which tells us that the tibble is grouped by two variables with four unique combinations of values.

We can then calculate the mean number of prey eaten by bats in each of the four sex and age combinations:

```
bats_num_prey_grouped %>%
  summarise(mean_num_prey = mean(num_prey))
```

```
## `summarise()` regrouping output by 'Sex' (override with '.groups' argument)

## # A tibble: 4 x 3
## # Groups:   Sex [2]
##   Sex    Age        mean_num_prey
##   <chr>  <chr>              <dbl>
## 1 Female Adult               4.09
## 2 Female Juvenile            3.86
## 3 Male   Adult               4.79
## 4 Male   Juvenile            4.78
```

Excellent! The first row shows us that the mean number of prey eaten by female adult bats is 4.09, the second that for female juvenile bats the mean is 3.86, and so on. There are four rows in total because there are four unique combinations of Sex and Age in the original data.

It is worth pointing out that we've used a convention above of creating *intermediate objects* such as bats_num_prey_grouped. We can (you can) instead do everything in a single pipeline. We'll remind you how to do this in the next chapter.

5.2.4 USING group_by WITH OTHER VERBS

The summarise function is the only **dplyr** verb we'll widely use with group_by in this book. However, many other **dplyr** functions also work with datasets grouped with group_by. When mutate or transmute is used with a grouped object they still add new variables, but now the calculations occur 'by group'. Here's an example using transmute, which creates new variables, but unlike mutate subsequently drops the existing ones:

```
centr_wingspan <- bats %>%
  group_by(Age) %>%
  transmute(wingspan_centred = Wingspan_mm -
              mean(Wingspan_mm, na.rm = TRUE))
```

In this example we calculate the 'group mean-centred' version of the wingspan_mm variable. The new wingspan_centred variable contains the difference between each observed wingspan and the mean of whichever age group that observation is in. In the fish-dietary-restriction Workflow Demonstration on the *Insights* companion website[2] is an example of how to use the do function with group_by. There it is used to efficiently and reliably calculate the growth rate of each individual fish in the dataset.

5.2.5 REMOVING GROUPING INFORMATION

On occasion it's necessary to remove grouping information from a dataset. This is most often required when working with 'pipes' (Section 6.1) when we need to revert back to operating on the whole dataset. The ungroup function removes grouping information:

```
bats_grouped %>%
  ungroup() %>%
  glimpse()
```

```
## Rows: 633
## Columns: 15
## $ Row_order    <dbl> 1, 2, 3, 4, 5, 6, 7, 8, 9, 10, 11, 12, 13, 14, 15, 16, ...
## $ Bat_ID       <dbl> 366, 366, 367, 367, 367, 367, 367, 367, 367, 367, 367, ...
## $ Age          <chr> "Adult", "Adult", "Adult", "Adult", "Adult", "Adult", "...
## $ Sex          <chr> "Female", "Female", "Female", "Female", "Female", "Fema...
## $ Sp._Nr.      <dbl> 52, 22, 64, 98, 114, 19, 2, 4, 70, 81, 29, 49, 19, 79, ...
## $ Species      <chr> "Ethmia bipunctella", "Bradycellus verbasci", "Hoplodri...
## $ Class        <chr> "Insecta", "Insecta", "Insecta", "Insecta", "Insecta", ...
## $ Order        <chr> "Lepidoptera", "Coleoptera", "Lepidoptera", "Lepidopter...
## $ Family       <chr> "Depressariidae", "Carabidae", "Noctuidae", "Noctuidae"...
## $ Pest         <chr> "no", "no", "no", "no", "yes", "yes", "no", "no", "yes"...
## $ Migratory    <chr> "no", "no", "no", "yes", "yes", "yes", "no", "no", "yes...
## $ Wingspan_mm  <dbl> 23.5, NA, 31.0, 52.5, 40.0, 37.5, NA, 20.5, 39.0, 29.0,...
## $ No._Reads    <dbl> 32672, 411, 14796, 8953, 3472, 3282, 2178, 1807, 1067, ...
## $ Date         <chr> "20.07.12", "20.07.12", "20.07.12", "20.07.12", "20.07....
## $ Date_proper  <date> 2012-07-20, 2012-07-20, 2012-07-20, 2012-07-20, 2012-0...
```

Looking at the top right of the printed summary, we can see that the Group: part is now gone.

---

[2] http://insightsfromdata.io

A few tips about naming the datasets and variables we create. Above, we started with `bats`, then created `bats_grouped`, then `bats_num_prey`, and then `bats_num_prey_grouped`. We created these new names quite carefully, to help us easily see where the datasets came from and what they contain. We could have called them `bats1`, `bats2`, `bats3`, but these names are less informative. We also created new variables within these datasets: we made a `num_prey` variable within the `bats_num_prey` tibble and a `mean_num_prey` variable (which we didn't assign to a new tibble). Does this seem confusing? If so, try to come up with your own naming conventions that work for you. We often use `bats` for the imported dataset, and then prefix any created datasets with `bats_` (this is just our convention; you could do something different). We strongly suggest you *do not* give variables within a dataset the same name as the dataset—use something like `bats` to differentiate datasets from variable names.

5.3 Summing up and looking forward

Isn't **dplyr** great?! You are now equipped to grow your R abilities very rapidly and efficiently. You will be able to do in a few lines what would have taken several tens of lines previously. We recently helped a colleague with several tens of lines of code do the same work with a couple of **dplyr** lines that were safe (it was not possible to make some of the errors that were previously possible), easier to read, and computationally faster. The only credit we can take is, however, that we took the time to learn **dplyr**, and perhaps even that is misplaced. All credit really goes to its creators, in particular the vision and leadership of Hadley Wickham.

Take your time to make sure you understand these **dplyr** functions. These are the functions (verbs) you will definitely continue to use as you explore your own data—they are the cornerstones of producing insights from tables (`group_by` and `summarise` in particular). And they are super-effective at creating summaries to use with **ggplot2** and your visualizations.

The next chapter will look at some additional R functions and operators that make tidying data and working with particular types of data easier. The chapter includes more detail on pipes as well as on managing strings, dates, and that age-old problem of changing data between long and wide formats.

Before then, go have some fun with some friends; try to resist preaching about how great R is.

Dealing with data 2

Expanding your toolkit

In this chapter we go through some miscellaneous but important topics, all of which you experienced in the Workflow Demonstration. We've isolated these because they are about programming and characteristics of data that often cause beginners and even experienced data analysts to lose sleep, cry, threaten their computers, or all of the above.

The topics are:

- *Pipes*: a mechanism for moving data from one operation to another. A bit like the conveyor belts that might connect the different machines on a production line. Materials (data) are processed a bit by one machine (function), and then travel on a conveyor belt (i.e. are piped) to the next machine (function), and so on.
- *Strings*: how words and text are represented in computers. Often ignored in introductory courses and can be very frustrating to deal with, but the **stringr** package takes the sting out of strings!
- *Dates and times*: until recently a proper pain in the **** in R, Excel, or anywhere on a computer. Now much simpler and straightforward to

Insights from data with R: An Introduction for the Life and Environmental Sciences. Owen L. Petchey, Andrew P. Beckerman, Natalie Cooper and Dylan Z. Childs, Oxford University Press (2021). © Owen L. Petchey, Andrew P. Beckerman, Natalie Cooper and Dylan Z. Childs. DOI: 10.1093/oso/9780198849810.003.0006

deal with, at least at the level most of us need to. Still frustrating, but the **lubridate** package is very good at managing dates.

- *Wide and long arrangement*: changing data from long to wide format, and from wide to long, used to be akin to a novice doing brain surgery. The `pivot_wider` and `pivot_longer` functions in the **tidyr** package are quite intuitive, such that we no longer always need to look up how to go from long to wide and wide to long.

If any of that did not make sense, read on and then come back to it!

6.1 Pipes and pipelines

As you have seen, we often use the various **dplyr** verbs (and other functions) in combinations, linked together by pipes. We use one function after another, e.g. first `mutate`, then `group_by`, and then `summarise` in a sequence. The pipe operator `%>%` allows us to make the logic of such sequences more transparent. And all of those advantages mean that once people start using pipes, they never look back.

6.1.1 WHY DO WE NEED PIPES?

Let's look again at our calculation in Chapter 5 of the mean number of prey of bats of each sex and age combination, but first do it *without* pipes:

```
bats_grouped <- bats %>%
  group_by(Bat_ID, Sex, Age)

bats_num_prey <- bats_grouped %>%
  summarise(num_prey = n())

bats_num_prey_grouped <- bats_num_prey %>%
  group_by(Sex, Age)

bats_num_prey_grouped %>%
summarise(mean_num_prey = mean(num_prey))
```

Seems fine, and it is. There's nothing wrong with this way of doing things. However, this approach to building up an analysis is quite verbose—especially if the analysis involves more than a couple of steps—because we make intermediate datasets that we never again need (e.g. `bats_grouped`, `bats_num_prey`, and `bats_num_prey_grouped`). It also tends to clutter R's memory with lots of data objects we don't need.

Here's the same thing done using pipes:

```
bats %>%
  group_by(., Bat_ID, Sex, Age) %>%
  summarise(., num_prey = n()) %>%
  group_by(., Sex, Age) %>%
  summarise(., mean_num_prey = mean(num_prey))
```

First, please check for yourself that the outcome is exactly the same. Combining several functions into a pipeline has the dual benefit of keeping our code concise and readable while avoiding the need to clutter the global environment with intermediate objects. This approach involves the 'pipe' operator, `%>%` (no spaces allowed). This isn't part of base R. Instead, it's part of a package called **magrittr**. But there's no need to install this if we're using **dplyr**, because **dplyr** makes `%>%` available to us.

The `%>%` operator has become very popular in recent years. The main reason for this is because it is super-fantastic. It allows us to specify a chain of function calls in a (reasonably) human-readable format. How so? Every time we see the `%>%` operator it means the following: take whatever is produced by the left-hand expression and use it as an argument in the function on the right-hand side. The `.` serves as a placeholder for the location of the corresponding argument. This means we can understand what a sequence of calculations is doing by reading from left to right, top to bottom, just as we would read the words in a book.

The example above takes the `bats` data, groups them by `Bat_ID`, `Sex`, and `Age`, then takes the resulting grouped tibble and applies the `summarise` function to it to calculate the number of prey using the `n()` function, then takes the result and groups it by `Sex` and `Age`, and then uses the `summarise` function to calculate the mean number of prey.

The n() function doesn't need any arguments. Perhaps because of that, and the fact that it is a function with a one-letter name, we often forget to use the brackets. This will cause a frustrating error. Get a piece of paper (at least A4/letter size) and write n() big enough to fill it. Stick the paper on your wall above your computer. Now, every time you forget the brackets, draw a little unhappy face on the paper. When you're done forgetting, send us your piece of paper.

When using the pipe operator, we can often leave out the . placeholder. This signifies which argument of the function on the right of %>% is associated with the result from the left of %>%. If we choose to leave out the ., the pipe operator assumes we meant to slot it into the *first* argument. This means we can simplify our example:

```
bats %>%
  group_by(Bat_ID, Sex, Age) %>%
  summarise(num_prey = n()) %>%
  group_by(Sex, Age) %>%
  summarise(mean_num_prey = mean(num_prey))
```

This is why the first argument of a **dplyr** verb is always the data object. This convention ensures that we can use %>% without explicitly specifying the argument to match against. As all **dplyr** verbs return a data frame or tibble, there is never an issue in piping from one operation to another!

Remember, R does not care about white space, which means we can break a chained set of function calls over several lines if it becomes too long, as we did above. In fact, people often place each part of a pipeline onto a new line to help with overall readability. It is also possible to add annotation in between pipes. Finally, when we need to assign the result of a chained function we have to break the left-to-right rule a bit, placing the assignment at the beginning:

```
new_data <- bats %>%
  # create group structure for summarising number of prey
  group_by(Bat_ID, Sex, Age) %>%
  summarise(num_prey = n()) %>%
  # create group structure for calculating mean number of prey
  group_by(Sex, Age) %>%
  summarise(mean_num_prey = mean(num_prey))
```

```
## `summarise()` regrouping output by 'Bat_ID', 'Sex'
(override with `.groups` argument)

## `summarise()` regrouping output by 'Sex' (override with
`.groups` argument)
```

```
new_data # see the output
```

```
## # A tibble: 4 x 3
## # Groups:    Sex [2]
##   Sex    Age       mean_num_prey
##   <chr>  <chr>             <dbl>
## 1 Female Adult              4.09
## 2 Female Juvenile           3.86
## 3 Male   Adult              4.79
## 4 Male   Juvenile           4.78
```

Why is %>% called the 'pipe' operator? The %>% operator takes the output from one function and 'pipes it' to another as the input. It's called 'the pipe' for the simple reason that it allows us to create an analysis 'pipeline' from a series of function calls. Incidentally, if you Google the phrase 'magritte pipe' you'll see that **magrittr** is a rather clever name for an R package.

If you remain unconvinced about piping, read on…

6.1.2 ON WHY YOU SHOULDN'T NEST FUNCTIONS

There's another way to use functions in combinations to avoid making intermediate datasets: by nesting functions within each other:

```
summarise(
  group_by(
    summarise(
      group_by(bats, Bat_ID, Sex, Age),
      num_prey = n()),
    Sex, Age),
  mean_num_prey = mean(num_prey))
```

This is equivalent to the previous example, but the functions are nested inside each other. We have placed the `group_by` function call inside the list of arguments to `summarise`, and again for the other `group_by` `summarise`. Note that we carefully used new lines, and RStudio indented the code to achieve maximum beauty and readability.

The trick is to remember that with nesting, you have to read from the inside outwards to understand what is going on. With the code above, because we added line breaks, one must start in the middle line and then read up and down at the same time!!! We get the same result without having to store intermediate data. However, there are a couple of very good reasons why this approach is not advised:

- It seems difficult to argue that nesting functions inside one another results in particularly intuitive code. Reading outwards (up and down simultaneously) from the inside of a large number of nested functions is really quite tricky.
- Even for experienced R users, using function nesting is a fairly error-prone approach. For example, it's very easy to accidentally put an argument or two on the wrong side of a closing) . If we're lucky, this will produce an error and we'll catch the problem. If we're not, we may just end up with nonsense in the output.

(Confession: We used to do quite a bit of nesting. We survived. We weren't sent to prison. We don't often use nesting now, however.)

Other types of pipe. The %>% pipes a dataset to the next function. We can pipe other things if needed (which may not be so often). If you'd like to learn more about other pipes, please take a look at the relevant material on the *Insights* companion website.[1]

6.2 Subduing the pesky string

Imagine that on the desk before you are many small wooden cubes, each with a letter of the alphabet printed on it and a hole through it. You also have a piece of string onto which you can thread some of the cubes. When you do so, you create a necklace or bracelet out of a 'string of letters', similar to a necklace made of a string of pearls or string of beads. This is one way to explain why we refer to a collection of letters as a *string*.

Put another way, datasets often contain words, and we call these words *strings*. Often these words are not quite as we would like them to be, and hence we need to manipulate them. Because words are such flexible things, there are endless types of manipulations we might want and/or need to make to strings.

In the first part of the Workflow Demonstration (Section 3.7.3), we manipulated the variable names (to make them easier to refer to/type) with the following two instructions:

```
names(bats)<- str_replace_all(names(bats), c(""="_"))
names(bats)<- str_replace_all(names(bats), c("\\(" ="","\\)"=""))
```

The function `str_replace_all` is from the **stringr** package, a package that contains many functions for manipulating strings. It is a good first stop for how to do manipulations on strings, and the associated cheat sheet is a great resource (Google 'cheat sheet stringr rstudio').

We also used a function `case_when` to replace words in a variable (in this case we were replacing abbreviated words with complete words):

[1] http://insightsfromdata.io

```
bats <- bats %>%
  mutate(
    Sex = case_when(Sex == "M" ~ "Male",
                    Sex == "F" ~ "Female"),
    Age = case_when(Age == "Ad" ~ "Adult",
                    Age == "Juv" ~ "Juvenile")
  )
```

Similarly, we may find that the strings in a variable contain two pieces of information, and that these two pieces of information are best separated into two variables so that we can use them more easily in an analysis. We can do this with the separate function from **tidyr**. Here's an example to show how it works. We'll start with some made-up data:

```
# first make an example dataset
df <- tibble(label = c(NA, "a-1", "a-2", "b-3"))
glimpse(df)
```

```
## Rows: 4
## Columns: 1
## $ label <chr> NA, "a-1", "a-2", "b-3"
```

Imagine that label contains information about an experiment, where the letter signifies the experimental 'treatment' (e.g. temperature, acidity) and the number corresponds to different 'replicates' (test tube, field plot). We might want to be able to identify cases according to the treatment they belong to, which means we need to split up the two bits of information in label:

```
df %>%
  separate(label, c("treatment", "replicate"), sep="-") %>%
  glimpse()
```

```
## Rows: 4
## Columns: 2
## $ treatment <chr> NA, "a", "a", "b"
## $ replicate <chr> NA, "1", "2", "3"
```

We started with one variable, `label`, and then split it into two variables, `treatment` and `replicate`, with the split made where - occurs. The `unite` function does the opposite, by the way.

Two additional things we find ourselves doing, especially when working with other people's data, include:

- The **stringr** functions `str_to_upper` and `str_to_lower` convert a string to upper and lower case. Standardizing cases can help us remember names, since we don't need to remember the capitalization. It can also increase correspondence of names among datasets: R will not match names that have everything in common except capitalization. Changing all names to lower case removes case differences as a possible reason for a lack of matches between names.

- We can trim leading and trailing white space with the **stringr** `str_trim` function. Spaces at the start or end of a value are problematic, because they create different values, e.g. `blue` and `blue` are different values as far as R is concerned, even though we, as clever humans, see those two strings as the same word. When working with R, it is generally undesirable to have two values for what should be one.

You can do a lot more than this with **stringr**. For example, you can use it to identify matches to a particular word, replace certain parts of a string, or extract bits of a string. If you do find yourself wrestling with messed-up strings, check out the **stringr** cheat sheet[2] or its website[3] and dive into the package. Trust us, you'll probably find you save yourself a whole lot of time in the long term.

Regular expressions are a powerful but rather impenetrable tool for working with strings. You may never need them (you won't in this book), but if you work with strings a lot, it will be worth checking

---

them out. There are lots of resources online to learn about them. One that we particularly like is **RVerbalExpressions**.[4]

As a quick example, one regular-expression function frequently used with **dplyr** is `grepl`. You can combine it with `filter` to subset a data frame with a 'string' of information, like subsetting on any family in the bat data that is a variation on *Noctuidae*, by specifying the string 'Noct':

```
bats %>%
  filter(grepl("Noct", Family)) %>%
  distinct(Family)
```

6.3 Elegantly managing dates and times

Dates and times have always been quite difficult to deal with in data analysis software. By 'dealing with' we mean getting the software to understand that each value is a point in time, and for it to know when that point in time is. Knowing this, the software can make calculations on date/time values, for example to find the number of days between two dates. It can also make a graph with dates/times on the *x*-axis, and plot values at the date/time they were recorded. (These benefits are demonstrated below.) Thus it's a good idea to appropriately deal with dates/times.

6.3.1 DATE/TIME FORMATS

One reason for dates/times being not such easy data to deal with is that there are many methods for writing down dates and times. We might have `13-10-2019` (day–month–year order), or `10-13-2019` (month–day–year order), or `2019-10-13` (year–month–day order). The order in which hours, minutes, and seconds are recorded is more standardized

---

[4] https://rverbalexpressions.netlify.com/

(hours–minutes–seconds), though sometimes the seconds are not given (14:31) and sometimes they are (14:31:00). Furthermore, times could be in 12 hour (2:00 PM) or 24 hour (14:00) format.

Such variability makes it difficult for us to tell our software how to 'read' (also known as 'parse') the date/time data. Is it in 12 or 24 hour time format? Does the year come first or last, and what about the day and month? Is the year written in full (2019) or only as the last two numbers (19)?

The **lubridate** package helps us read read date/time information in R. When using its functions for parsing, we often only need to know the *order* in which day, month, and year, and hours, minutes, and seconds are recorded in the data.

6.3.2 DATES IN THE BAT PROJECT DATA

Let us look again at the bat data (make sure you are in the bat project and working in an appropriate script file and location in that file). If required, load libraries, read in the data, and view the date with `glimpse`:

```
bats <- read_csv("data/bat_sex_diet_Mata_etal_2016.csv",na ="na")
```

```
bats %>%
  select(Date) %>%
  glimpse()
```

```
## Rows: 633
## Columns: 1
## $ Date <chr> "20.07.12", "20.07.12", "20.07.12",
    "20.07.12", "20.07.12", "2...
```

We see that the dates when a sample was collected are in the variable named `Date` and are, at the moment, strings (`<chr>`). And we see that the dates are recorded as `dd.mm.yy`, where `dd` is the day, `mm` is the month, and `yy` is the year. The next step is to 'parse' the dates to turn them into proper dates in R.

We have the option of using the following functions in the **lubridate** package: `ymd`, `ydm`, `mdy`, `myd`, `dmy`, or `dym`. Our choice should

correspond with the order of day, month, and year in our data. Hence
we use the dmy function:

```
bats <- bats %>%
  mutate(Date_proper = dmy(Date))
```

The dmy function does not return any warnings, so it seems all the dates
have been parsed. If we look at the dataset in R now, we see a new variable
Date_proper and that it is a <Date> formatted variable.

When attempting to parse dates and/or times we sometimes get a
warning that some dates could not be parsed. One possible cause is
that the dates have been entered into the raw data in inconsistent
patterns, e.g. sometimes with year first and sometimes with day first.
Another, though related, reason is that we might have told R to expect
a month, yet it saw a number greater than 12.

6.3.3 WHY PARSE DATES?

When it is read by read_csv, we see the date variable in our bat dataset is
a character. It is just words. The values bear no time-related relationship to
each other. R does not know that 01.01.2019 is two days before 03.01.2019,
for example. R would not be able to make a time series of values with dates
and/or times on an axis. Take a look at this example, in which we ask R for
the minimum and maximum of the unparsed date variable:

```
bats %>%
  summarise(min_date = min(Date),
            max_date = max(Date))
```

```
## # A tibble: 1 x 2
##    min_date max_date
##    <chr>    <chr>
## 1 03.08.13 24.09.13
```

And compare this with when we ask the same of the parsed date variable:

```
bats %>%
  summarise(min_date = min(Date_proper),
            max_date = max(Date_proper))
```

```
## # A tibble: 1 x 2
##   min_date   max_date
##   <date>     <date>
## 1 2012-07-20 2013-10-16
```

In the first case, 03.08.13 comes first because R sorts the character variable that is Date alphabetically, by the first character, then the second, then the third, and so on. In the second case, the minimum is the earliest date in the variable, and the maximum is the last. The minimum and maximum are appropriate. Try plotting various graphs and mapping either Date or Date_proper onto the *x*-axis.

6.3.4 MORE ABOUT PARSING DATES/TIMES

Let's look a bit more closely at the parsing of dates, by creating one date, showing it, and then parsing it:

```
the_date <- "13.1.19"
the_date
```

```
## [1] "13.1.19"
```

```
dmy(the_date)
```

```
## [1] "2019-01-13"
```

We created a string in the object the_date. The string was 13.1.19. We then printed this—R has no idea it's a date—the string appears exactly as when we entered it. But look at dmy(the_date) —R tells us this is 2019-01-13: year (all four numbers), followed by month (always two digits), followed by day (always two digits), each separated by a -.

Year, then month, then day is often used to represent dates, because then one can easily sort them into ascending or descending order.

The **lubridate** package also has functions for working with times:

```
the_time <- "20:10:35"
the_time
```

```
## [1] "20:10:35"
```

```
hms(the_time)
```

```
## [1] "20H 10M 35S"
```

We made a string in the object the_time and showed it. Again, it's just as we typed it in (20:10:35). We parse this using the hms function and then see that we get 20H 10M 35S. Excellent. R knows it's a time.

And the same for dates *and* times, using the dmy_hms function to do the parsing:

```
the_date_time <- "13.1.19 20:10:35"
the_date_time
```

```
## [1] "13.1.19 20:10:35"
```

```
dmy_hms(the_date_time, tz = "CET")
```

```
## [1] "2019-01-13 20:10:35 CET"
```

Now R tells us that it understands the entered date and time to be 2019-01-13 20:10:35 CET. We set the time zone to be Central European Time, with tz = "CET".

6.3.5 CALCULATIONS WITH DATES/TIMES

There are many useful **lubridate** functions for working with dates and times. Here are some for extracting parts of dates and times. (We will not explain them. We think they are relatively self-explanatory, and also wish to encourage you to find things out for yourself.)

```
parsed_the_date_time <- dmy_hms(the_date_time, tz = "CET")
second(parsed_the_date_time)
```

```
## [1] 35
```

```
minute(parsed_the_date_time)
```

```
## [1] 10
```

```
hour(parsed_the_date_time)
```

```
## [1] 20
```

```
day(parsed_the_date_time)
```

```
## [1] 13
```

```
wday(parsed_the_date_time, label = TRUE)
```

```
## [1] Sun
## Levels: Sun < Mon < Tue < Wed < Thu < Fri < Sat
```

```
week(parsed_the_date_time)
```

```
## [1] 2
```

```
month(parsed_the_date_time, label = TRUE)
```

```
## [1] Jan
## 12 Levels: Jan < Feb < Mar < Apr < May < Jun < Jul < Aug <
      Sep < ... < Dec
```

```
year(parsed_the_date_time)
```

```
## [1] 2019
```

There are also functions for rounding. You can very likely work out for yourself what these do:

```
floor_date(parsed_the_date_time, unit = "years")
```

```
## [1] "2019-01-01 CET"
```

```
round_date(parsed_the_date_time, unit = "day")
```

```
## [1] "2019-01-14 CET"
```

With the following operation, we can make a new variable containing the time since the first sampling date. In this case the units chosen are days:

```
bats <- bats %>%
  mutate(time_since_first_sample = Date_proper - min(Date_proper))
```

We can also construct intervals between two dates/times, get the duration of such an interval, and set the units of time. Here we construct the interval from the first date of the study to the last, using the `interval` function:

```
study_interval <- interval(ymd("2012-07-20"), ymd("2013-10-16"))
study_interval
```

```
## [1] 2012-07-20 UTC--2013-10-16 UTC
```

Then we get the duration of this interval with the function `as.duration`:

```
study_duration <- as.duration(study_interval)
study_duration
```

```
## [1] "39139200s (~1.24 years)"
```

Cool! We get told the duration of the interval in seconds, and in approximate years. If we then divide by the number of seconds in one day (which we can get with the code `ddays(1)`), we get the number of days of the study. We can do the same for weeks or years:

```
study_duration/ddays(1)
```

```
## [1] 453
```

```
study_duration/dweeks(1)
```

```
## [1] 64.71429
```

```
study_duration/dyears(1)
```

```
## [1] 1.240246
```

```
study_duration/dmonths(1)
```

```
## [1] 14.88296
```

We should wonder about the `dmonths` function, because months vary in their duration. If you run `dmonths(1)`, you will see that the duration is `2629800s (~4.35 weeks)`. Have a think about how **lubridate** calculates this duration of a month. This might make us wonder about the duration of `dyears(1)`—probably it's the duration of a normal year and not that of a leap year. Let's check:

```
dyears(1)/ddays(1)
```

```
## [1] 365.25
```

So the **lubridate** package assumes 365.25 days in a year, so it's the average length of a year. This highlights, that we need to be aware of these and other intricacies when working with dates.

There are lots of other things we can do with dates. As with many of the great packages, there is a cheat sheet[5] for **lubridate**. There are also many nice web pages.[6] The bottom line is that we should always make our dates into proper dates in R and use the nice functionality of **lubridate** to work with them.

 When we parsed the date variable, we converted its 'type'. We often see and use functions that convert one type of variable into another. These functions usually start with `as.`, like we just saw with `as.duration`. Other examples are `as.numeric`, `as.character`, `as.factor`, and so on. Before you try one, have a think about their likely limits. We might realize, for example, that it could be tough for R to convert a character to a numeric (e.g. converting the string `R is awesome` into a number is going to fail):

6.4 Changing between wider and longer data arrangements

We have gone on a bit about data being 'tidy'. Truth is, though, untidiness is sometimes not so bad. There is, in fact, a type of untidiness called wide-format data that can be valuable, in contrast to the long format of classically tidy data. It is useful, therefore, to be able to change data back and forth between tall and wide data as and when we need.

---

[5] https://github.com/rstudio/cheatsheets/raw/master/lubridate.pdf
[6] https://lubridate.tidyverse.org/

For example, if we wished to graph time series (some value on the *y*-axis and time on the *x*-axis), we would be happy with tidy data. But if we wished to plot values from one year against values from the previous year, then we would be happy with the data being 'untidy' (e.g. data from different years spread across multiple columns—the wide format). Thankfully, it isn't too tricky to switch data from wider to longer arrangement, or from longer to wider arrangement.

6.4.1 GOING LONGER

Some of the early insights during the bat Workflow Demonstration were about the distribution of the three response variables, which corresponded to the three questions we sought to answer. The first three sections of Section 4.4 involved us plotting these three distributions, one for each of the number of prey, the mean wingspan of prey, and the proportion of migratory prey. We made each of these three histograms with a separate ggplot.

There is, however, a rather more efficient method to make all three histograms. The method involves first rearranging the data, and then plotting each of the different response variables in a different facet. As well as learning this specific method, we will here learn generally about how to gather data from multiple columns into one column (i.e. go from wider to longer data), and how to spread data from one column across multiple columns (i.e. go from longer to wider data).

The dataset we use is `prey_stats`, so please ensure you have it available in R. First, be very clear in your head about how the data are arranged therein. The dataset has 142 rows and 6 variables. Three of those variables are our three response variables: `num_prey`, `mean_wingspan`, and `prop_migratory`.

The data for these three variables are contained in three columns in the dataset. In order to make all three histograms with one ggplot, we need to pivot the values of the three variables into one column (let's call it `response_value`), and then make a new column (let's call it `response_variable`) in which the values are the names of original variables that were in the columns. The **tidyr** package provides the `pivot_longer` function to do this, with arguments `names_to`, which defines the new variable made up of the original variable names, and `values_to`, which collects the values from those variables. (Note that if we gather the data from three columns, each with 142 values, into one column, then that new column should have 142 × 3 = 426 entries.)

That is probably quite confusing. Let's do it and show the result…

```
longer_prey_stats <- prey_stats %>%
  pivot_longer(names_to = "response_variable",
               values_to = "response_value",
               cols = c("num_prey",
                        "mean_wingspan",
                        "prop_migratory"))
longer_prey_stats
```

```
## # A tibble: 426 x 5
## # Groups:    Bat_ID, Sex [142]
##    Bat_ID Sex     Age    response_variable response_value
##     <dbl> <chr>   <chr>  <chr>                      <dbl>
## 1     366 Female  Adult  num_prey                       2
## 2     366 Female  Adult  mean_wingspan               23.5
## 3     366 Female  Adult  prop_migratory                 0
## 4     367 Female  Adult  num_prey                       9
## 5     367 Female  Adult  mean_wingspan               35.6
## # ... with 421 more rows
```

Good, our new dataset has 426 rows as expected. It still has the first three columns (`Bat_ID`, `Sex`, and `Age`), but instead of three columns after that (one for each response variable) we have two columns: a column named `response_variable` and a column named `response_value`. Open the dataset with `View(longer_prey_stats)` and scroll down; you will see there are 142 rows when `response_variable` is `num_prey`, 142 when `response_variable` is `mean_wingspan`, and 142 when `response_variable` is `prop_migratory`.

How did `pivot_longer` do its magic? Or rather, how did we tell it what to do? First, we piped the `prey_stats` dataset into it. Then

we came to name-value pair arguments: `names_to = "response_`
`variable"` and `values_to = "response_value"`. The first gives
the name of the new column that will contain what were the names of the
columns in `prey_stats` across which the data are spread. The second
gives the name of the new column that will contain the values. The final
argument (`cols = ...`) gives the names of the three variables that we
want to take the values from. This could still be confusing, so let us consider
another example.

In case it's useful for understanding what happened, and the new dataset,
let us put the three histograms in one ggplot, with a facet for each response
variable (the resulting graph is in Figure 6.1):

To make a facet for each response variable, we used `facet_wrap( ~`
`response_variable, scales = "free")`. This says 'make a facet
for each of the values in the `response_variable` variable, and make

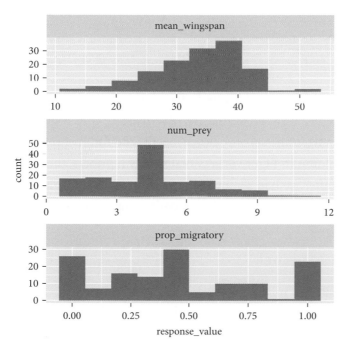

Figure 6.1 A histogram for each of the response variables, obtained by first
making the data longer and then faceting. See the main text for explanation.

the scales of each of the facets *free* of each other' (i.e. do not use the same *x* and *y* limits/ranges for all facets). Experiment by removing `scales =` `"free"` for clarity about why we needed to do this.

In case you have uncertainty about making data longer, let's try a different way of learning. Imagine a dataset in which data for each of several years are in a separate column—data for different years are spread across columns—and each row contains lots of observations, each from a different year. This might be, for example, the `Amount` of some `Item` (e.g. a particular food item such as milk) in a particular year. The dataset might then look like that shown in Figure 6.2b. Each item has one row of data, and has multiple columns, one for each year. This arrangement of the data is often known as wide format (and it is not tidy). To make these data tidy, we need to collect the data that are spread across the year columns into one column. The pseudocode to do this would be

```
food_item_cost %>%
  pivot_longer(names_to = "Year",
               values_to = "Amount",
               2:5)
```

This `pivot_longer` would change the data into the arrangement given in Figure 6.2a. Note that we have used `2:5` as a quick way to say 'gather data from the second, third, fourth, and fifth columns'. By the way, the Year column in Figure 6.2a contains only numbers, though this would only be so after we had removed the 'Y' (e.g. `Y1981`). This would require some string manipulation and then conversion to numeric type (give it a go, now or later).

6.4.2 GOING WIDER

The opposite of making data longer is to spread the data from one column into multiple columns. The function to do this is, unsurprisingly, `pivot_wider`. Here is how we can spread the gathered version (`longer_prey_stats`) to get back to the arrangement of the original `prey_stats`:

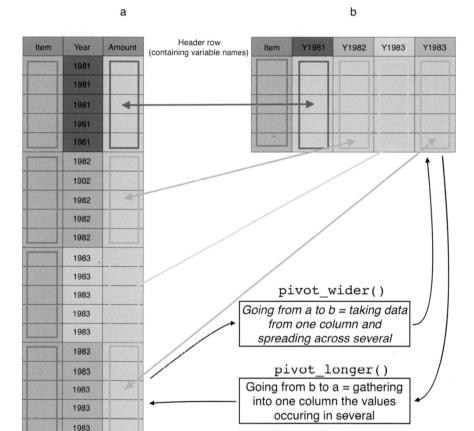

Figure 6.2 A schematic of a dataset (a) in long/tidy format and (b) in wide format, with colours and arrows linking the same data in the two different formats.

```
longer_prey_stats %>%
  pivot_wider(names_from = "response_variable",
        values_from = "response_value")
```

```
## # A tibble: 142 x 6
## # Groups:   Bat_ID, Sex [142]
##   Bat_ID Sex    Age   num_prey mean_wingspan prop_migratory
##    <dbl> <chr> <chr>    <dbl>         <dbl>          <dbl>
## 1    366 Female Adult      2          23.5            0
## 2    367 Female Adult      9          35.6            0.556
## 3    584 Female Adult      6          36.6            0.5
```

```
## 4      598 Male    Adult       5          39.2        1
## 5      606 Male    Adult       6          18.7        0
## #  ... with 137 more rows
```

That was rather easy, huh? We tell it the column in which the names of the new columns come from with the `names_from` argument, and the column containing the data to be spread across multiple columns with the `values_from` argument.

 The tidyverse evolving. The functions `pivot_longer` and `pivot_wider` were developed quite recently. Before then, the functions to use were called `gather` and `spread`. Note that the meaning of 'pivot' in these **tidyr** functions is different to what is meant when used in the context of *pivot tables* in Excel.

6.5 Summing up and looking forward

Remember that at the beginning of this chapter we summarized what you would experience, and said that if you then did not understand the summary to come back after you had read the chapter? Please go back to the beginning of the chapter and look at the list of R topics and what we wrote about what you would learn. We hope you now fully understand the value of pipes, how to work efficiently and reliably with strings and dates, and how to readily change data between wide and long formats. And that you understand why you should not do some things you might encounter others doing!

As always, take a good break before you embark on the next chapter, which is a dive into the wonders of **ggplot2** for making figures/graphs. Also, take a moment to review your progress so far, acknowledging how much you have learned, and exposing where you think you might need to spend a little more time.

Now might also be a good time to look up your local R user group. Or if you don't have one close to you, maybe start one? They can be a bit daunting—it's common to think 'surely everyone is an R master and novices like me will be shunned!'—but the reality could not be further from the truth. They are typically very welcoming to new converts, or even sceptics, though think twice about entering the church if you're not ready to be preached to! A brilliant example is the Coding Club, led by students at Edinburgh University. They have resources you can use to set up your own group on their website.[7]

---

[7] https://ourcodingclub.github.io/

Getting to grips with ggplot2

Making graphs is one of the most fundamental activities in the quest for insights from data. A good graph communicates the design of your study/experiment and gives yourself and other viewers rapid access to patterns in the data that are associated with specific questions. If you think back to all of the steps in our workflow diagram (Chapter 2), one of the first steps is making a *sketch* of what you expect to find in the data you collect. Actually making those graphs, with your data, sharing them, and communicating your science is one of the best feelings.

We have made quite a few graphs already, but only briefly explained how `ggplot` works. We need a deeper understanding... hence this chapter. We focus on making graphs with the **ggplot2** package. The **ggplot2** package seems to sit in a 'sweet spot' between simplicity and complexity. It can be quite simple, but also very powerful. It can help us to produce quite complex visualizations, with elements such as graphical keys, without the need to write lines and lines of R code. It is, nevertheless, still flexible enough for us to tweak the appearance of a figure so that it meets our specific needs. The **ggplot2** package was developed by Hadley Wickham to implement some of the ideas in a book called *The Grammar of Graphics* (hence 'gg'), by Leland Wilkinson (Springer, 2005).

Insights from data with R: An Introduction for the Life and Environmental Sciences. Owen L. Petchey, Andrew P. Beckerman, Natalie Cooper and Dylan Z. Childs, Oxford University Press (2021). © Owen L. Petchey, Andrew P. Beckerman, Natalie Cooper and Dylan Z. Childs. DOI: 10.1093/oso/9780198849810.003.0007

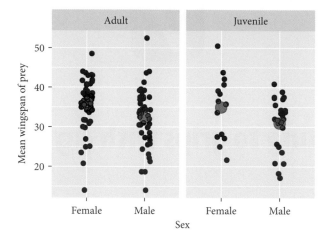

Figure 7.1 A recap of the graph we previously made in the Workflow Demonstration. We will use this graph to explain some of the principles of ggplot. (The graph shows the average moth size in the diets of male and female bats of two ages. Black dots show the average prey size of individual bats. Blue dots show averages for each sex and age combination.)

We have already looked at code to make graphs with ggplot during the Workflow Demonstration. Let's look at it again here, with the aim being to explore and more deeply explain some of the principles of the grammar used. We'll again use the bat diet data and will work through creating a graph we already saw, which is now presented again (Figure 7.1). It is a scatterplot. This is one of the most commonly used visualizations. It's designed to show how one numeric variable is related to another numeric or categorical variable.

To work along with us you will need to have made two datasets: prey_stats and mean_prey_stats. You can find how to do this in Section 4.4.6 in the Workflow Demonstration.

7.1 Anatomy of a ggplot

Learning some of the principles of ggplot can be helped by seeing a ggplot made in a way we rarely ever use and don't recommend for everyday use. It is long-winded but therefore very transparent. The method is shown in Figure 7.2: here is how that ggplot is composed:

1. Start with `ggplot()` to tell it we're beginning a new ggplot.
2. Define the 'scales' of the graph, e.g. the scaling of the *x*- and the *y*-axis.
3. Define the coordinate system, e.g. Cartesian coordinates.
4. Define any 'facets' of the graph. Facets are when the graph has multiple panels.
5. Specify any other features of the graph, e.g. axis labels.
6. Add two 'layers' of data, each with layer-specific information.

Notice that the first five points are specifying things about the whole graph, while the layers are made with information (e.g. data) specific to them. We have colour-coded the former in green and the latter in orange in Figure 7.2. Let's take a closer look at these elements, starting with the layers of data.

7.1.1 LAYERS

The code we're using as an example (Figure 7.2) contains two layers, one of black dots showing the mean wingspan of the prey consumed by each individual bat, and one of two blue dots showing the mean wingspan of the prey (across bats) for each of the four sex-by-age combinations. We make each layer with the `layer` function, and give six components:

- The *data*. Every layer needs some data, and **ggplot2** layers always and only accept data in one format, an R data frame (or tibble). Each layer can be associated with its own dataset. In our example (Figure 7.2), one layer is given `prey_stats` and the other is given `mean_prey_stats`.
- A *geometric object*, called a 'geom'. The geom tells **ggplot2** how to represent the layer—geoms refer to the objects we can actually see on

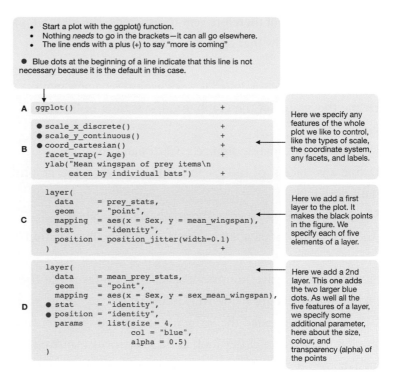

Figure 7.2 We show this method as it clearly shows the components of a ggplot (or any graph for that matter). Compare the code here with that in Figure 7.3, which shows a more compact and standard method for making the same figure. All explanations are in the text of the figure or the main text.

a plot, such as points, lines, or bars. In our example (Figure 7.2), both layers make points on the graph, since we wrote geom = "point".

- A set of *aesthetic mappings*, captured by aes. These describe how variables in the data are associated with the aesthetic properties of the layer. Commonly encountered aesthetic properties include the position (*x* and *y* locations), colour, and size of the objects (e.g. points) on a plot. Each layer can be associated with its own unique aesthetic mappings. Notice that an aesthetic mapping is defined by one or more name-value pairs, specified as arguments of the aes function. The names on the left-hand side of each = refer to the properties of our graphical object (e.g. the *x* and *y* positions). The values on the right-hand side refer to variable names in the dataset that we want to associate with these properties. Each geom only works with a particular set of aesthetic mappings. For example (Figure 7.2), `geom_point` at least needs aesthetic mappings for the *x* and *y* positions of the points (i.e. `aes(x = Sex, y = mean_wingspan)`). And `geom_histogram` needs only an *x*-position mapping.

- A *stat*. These transform the raw data in the data frame in some hopefully useful way. A stat allows us *within ggplot code* to produce summaries of our raw data. We rarely use these directly; they are most often automatically used when we ask for a geom that requires them (e.g. `geom_histogram` automatically uses `stat_bin`). The stat facility is one of the things that makes **ggplot2** particularly useful for exploratory data analysis. In our example (Figure 7.2), both layers have `stat = "identity"`, which means 'please don't make any transformation of the data before plotting them—just plot the data we supplied'.

- A *position adjustment*. These apply tweaks to the position of layer elements. They are most often used in plots like bar plots, where we need to define how the bars are plotted, but they can occasionally be useful in other kinds of plot. In our example (Figure 7.2), we jittered the points in the first layer so fewer of them would obscure each other, and did not adjust the position of points in the second layer (`position = "identity"`, where again `identity` means 'please make no changes to the position').

- Layer-specific *params* (parameters). These are a list of features of a layer, for example the colour of the data points, or the symbol used to show all data points. In our example (Figure 7.2), we used `params = list(size = 4, col = "blue", alpha = 0.5)` to say that all data points in that layer should be four times the default size, should be blue, and should be somewhat transparent. More about these layer-specific properties in Section 7.3.1.

7.1.2 SCALES

The scale part of a **ggplot2** object controls how the data are mapped to the aesthetic attributes. A scale takes the data and converts them into something we perceive in the graph, such as an *x*/*y* location, or the colour and size of points in a plot. A scale must be defined for every aesthetic in a plot. It doesn't make sense to define an aesthetic mapping without a scale, because there is no way for **ggplot2** to know how to go from the data to the aesthetics without one.

Scales operate in the same way on any every layer: they all have to use the same scale for the shared aesthetic mappings. This behaviour is sensible because it ensures that the information that is displayed is consistent.

In our example (Figure 7.2), we specified a discrete *x*-scale (`scale_x_discrete`), since the `Sex` variable that we mapped to the *x*-axis contains categories, whereas we specified a continuous *y*-scale, since the wingspan variables mapped to the *y*-axis are continuous (in both layers) (`scale_y_continuous`). Note that these functions that we used to specify the scales have lots of different options we can specify; e.g. try `scale_y_continuous(position = "right")` or `scale_y_continuous(transform = "log10")`.

7.1.3 COORDINATE SYSTEM

A **ggplot2** coordinate system takes the positions of objects (e.g. points and lines) and maps them onto the 2D plane that a plot lives on. You are probably familiar with the most common coordinate system: the Cartesian coordinate system is the one we've all been using ever since we first constructed a graph with paper and pencil at school. Most graphs use this coordinate system, so we won't consider any others (e.g. polar coordinates) in this book. In our example (Figure 7.2), we specified the Cartesian coordinate system (`coord_cartesian`).

7.1.4 FANTASTIC FACETING

A graph may contain only one facet, i.e. panel. But often we benefit from having separate panels for subsets of data, such as a panel for each sex or age. Facets help represent the structure of our study and emphasize patterns in our data. In our example (Figure 7.2), we asked for a separate facet for each of the values in the Sex variable by writing `facet_wrap ( ~ Age)`. The result is a multipanel plot, where each panel shares the same layers, scales, etc. The data plotted are the only thing that varies from panel to panel in this example.

Faceting is a *very* powerful tool that allows us to slice up our data in different ways and really understand the relationships between different variables. Together with aesthetic mappings, faceting allows us to summarize relationships among four to six variables in a single plot. That is, we can view four to six dimensions, which is pretty cool! (And this does not require 3D graphs.)

7.2 Putting it into practice

That was an example to show the anatomy of a ggplot, used to reveal how the grammar works, rather than for its efficiency. A more concise, standard approach to using **ggplot2** relies on the built-in default behaviours. We do not need to specify the scales, since `ggplot` can try to guess them from the variable types in the aesthetic mappings. We need not specify the coordinate system, since Cartesian is the default. Also, when we are using `geom_point` and would like `position = "identity"` or `stat = "identity"` we need not specify them, since these are the defaults for this geom. Finally, we usually do not use the `layer` function, but rather use a convenience function specific to the geom we want; e.g. in this case we can use `geom_point`, and then we do not need to specify `geom = "point"`.

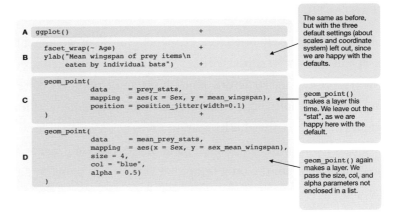

Figure 7.3 A more compact and standard method for making the same figure as was coded in Figure 7.2. All explanation is in the text of the figure or the main text.

After taking advantage of the defaults and the convenience functions (e.g. `geom_point`), we get the more compact and efficient code shown in Figure 7.3. This looks just like how we have previously used `ggplot`.

The **ggplot2** package is quite flexible, which means we can arrive at a particular visualization in a number of different ways. To keep life simple, we use one throughout this book, in which we specify each layer (geom) and its data separately, and never put any arguments inside the `ggplot` brackets.

7.2.1 INHERITING DATA AND AESTHETICS FROM `ggplot`

You won't have to look far outside this book, however, to find other ways of using `ggplot`, and very often with arguments in the `ggplot` function brackets. When you put the `data` and `aes` arguments in the `ggplot()` brackets, subsequent layers 'inherit' this information, unless you deliberately specify a change.

In the examples above we specified the data and aesthetics inside the geom function, like this for example:

```
ggplot() +
  geom_beeswarm(data = prey_stats,
                mapping = aes(x = Sex, y = mean_wingspan))
```

We can, however, specify them inside ggplot, like this:

```
ggplot(data=prey_stats, mapping =aes(x=Sex, y=mean_wingspan))+
  geom_beeswarm()
```

In this case, the geom 'inherits' the data and aesthetics from the ggplot function. This inheritance will always occur if it can (i.e. if ggplot has been given something to pass on, and if nothing is specified directly in the geom).

Inheritance can be quite useful when making multiple layers using the same data and aesthetics, or parts of aesthetics. Figure 7.4 is made with the following code, which uses such inheritance (and note that we could also have piped prey_stats into ggplot, as we have done in some previous examples):

```
ggplot(data = prey_stats,
       mapping = aes(x = Sex, y = mean_wingspan)) +
  geom_beeswarm() +
  geom_beeswarm(col - "red",
                size - 0.5)
```

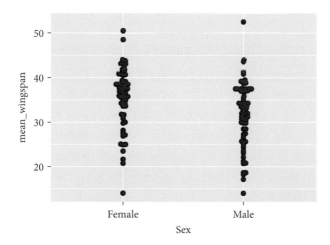

Figure 7.4 Adding two layers with geom_beeswarm, one with larger black dots, and then one with smaller red ones. In the code, both of the geom_beeswarm calls 'inherit' the same data and aesthetics from ggplot.

The following code will not work properly (it does not plot the layer of smaller red dots), because no data or aesthetics are specified in the second geom_beeswarm call, and there is nothing to inherit from the ggplot call. So, the second geom_beeswarm layer has no data or aesthetic mappings to work with. Try this yourself:

```
ggplot() +
  geom_beeswarm(data = prey_stats,
                 mapping = aes(x = Sex, y = mean_wingspan)) +
  geom_beeswarm(col = "red",
                 size = 0.5)
```

Which method should you use? Give the data and aesthetic mappings in the ggplot call or in each geom_ layer? For consistency, we will continue to give the aesthetic mappings in each layer, and we will pipe the dataset into ggplot. When doing our own work we tend to be less consistent, just doing whichever seems more sensible at the time, or perhaps just flipping a mental coin. We are, of course, always very aware of which we are doing. Always....

7.3 Beautifying plots

The default formatting used by **ggplot2** is generally fine for exploratory purposes. In fact, although they aren't universally popular, the defaults are carefully chosen to ensure that the information in a plot is easy to discern. These choices are a little unconventional, though. For example, published figures usually use a white background. For this reason, we often need to change the appearance of a plot once we're ready to include it in a report.

Our aim in this section, and throughout the book, is to learn a little bit about the underlying logic of how to customize **ggplot2**. We aren't going to attempt to cover the many different permutations, as what you will want to customize and how is quite personal. And there are many very nice sources that have been written about customizing ggplots, and also about important design issues, like choosing appropriate colours (think about people with colour vision deficiencies) and colour scales, and many other common design pitfalls. *The Fundamentals of Data Visualization*[1] by Claus O. Wilke is a very nice source.

Instead, we'll explain a bit of what is possible by reference to the small amount of beautification we did for Figure 7.1, and some of the main principles underlying the different routes to customization.

7.3.1 WORKING WITH LAYER-SPECIFIC GEOM PROPERTIES

What do we do if we need to change the properties of a geom? We're using the point geom at the moment. How might we change the colour or size of points in our scatterplot? It's quite intuitive—we set the appropriate arguments in the `geom_point` layer. We did this in our example code in Figure 7.3.

The point colour is set with the `colour` argument. There are many ways to specify colours in R, but if we only need to specify a few the simplest is to use a name R recognizes. The point size is specified with the `size` argument. The baseline is 1, and so here we increased the point size by assigning this a value of 1.5. Finally, we made the points somewhat transparent by setting the value of the `alpha` argument to be less than 1. In graphical systems the 'alpha channel' essentially specifies the transparency of something—a value of 0 is taken to mean 'completely invisible' and a value of 1 means 'completely opaque'.

Built-in colours in R. There is nothing special about 'blue' other than the fact that it is a colour name 'known' to R. There are over 650 colour names built into R. To see them, we can use a function called `colours` to print these to the Console. Try it (type `colours()` into the console in RStudio). The **colorspace** add-on package[2] contains comprehensive tools for working with colours in R.

---

[1] https://clauswilke.com/dataviz/
[2] http://colorspace.r-forge.r-project.org/

There are other arguments—such as `fill` and `shape`—that can be used to adjust the points. We won't look at these here. The best way to learn how these work is to simply experiment with them. You might try using `geom_point(shape = 21, colour = "purple", fill = "pink")` for fun....

The key message to take away from this little customization example is this: if we want to set the properties of a geom in a particular layer, we do so by specifying the appropriate arguments in the `geom_NAME` function that defines that layer.

When `fill`, `shape`, and `colour` are *not* encased in `aes()`, this applies the customization to all the data. However, as we saw in the Work-flow Demonstration, we can also specify `fill`, `shape`, and `colour` in the `aes` argument too. This generates a mapping between `fill`, `shape`, and `colour` and the levels of some categorical variable.

For example, in Chapter 4 we built a figure of the frequencies with which prey species appeared in the diets of bats, for only the 10 species with the greatest difference between male and female bats (Figure 4.13). This figure specified that the filling colours of `geom_col` should be mapped to the levels/categories in the `Sex` variable. This feature of **ggplot2**, alongside facets, is one of the most valuable tricksets for visualizing data and rapidly colouring or shaping points to reveal structure and pattern. A side effect of using `fill`, `shape`, and `colour` in the `aes` argument is the automatic creation of a legend/key.

*How should we arrange **ggplot2** code?* Take another look at the example. Notice that we split the code over three lines, placing each function on its own line. The whole thing will still be treated as a single expression when we do this because each line, apart from the last one, ends in a +. R doesn't care about white space. As long as we leave the + at the end of each line, R will consider each new line to be part of the same definition. Splitting the different parts of a graphical object definition across lines like this is a very good idea. It makes everything more readable and helps us spot errors. This way of formatting **ggplot2** code is pretty much essential once we start working with complex plots.

7.3.2 ADDING TITLES AND LABELS

What else might we like to tweak? Axis labels often need tweaking, since if we do nothing they are just the names of the data variables used to define the aesthetic mapping. The axis labels are a feature of the whole plot. They do not belong to a particular layer. This is why we don't alter axis labels by passing arguments to a function that builds a layer (`geom_point` in this case). Instead, we use the `xlab` and `ylab` functions to set the *x* and *y* labels, respectively, using + to add them to our graphical object. If we need to add a title to a graph, we can use the `ggtitle` function in the same way.

The `labs` function provides a more flexible alternative to `xlab` and `ylab`. It's more flexible because `labs` can be used to change the label of every aesthetic in the plot. For example, if we want to set the labels of the *x*- and *y*-axes and the label associated with `Age`, we can use this (where . . . refers to all the other required parts of the ggplot):

```
... +
  labs(x = "Sex of bat",
       y = "Mean wingspan of\nprey eaten by moth",
       col = "Bat age")
```

7.3.3 THEMES

The final route to customization we'll consider concerns something called the 'theme' of a plot. We haven't considered the **ggplot2** theme system at all yet. In simple terms, **ggplot2** themes deal with the visual aspects of a plot that aren't directly handled by adjusting geom properties or scales, i.e. the 'non-data' parts of a plot. This includes features such as the colour of the plotting region and the grid lines, whether or not those grid lines are even displayed, the position of labels, the font used in labels, and so on.

The **ggplot2** theme system is extremely powerful. Once we know how to use it, we can set up a custom theme to meet our requirements and then apply it as needed with very little effort. However, it's not an entirely trivial thing to learn about, because there are so many components of every plot. Fortunately there are a range of themes built into **ggplot2** that are easily good enough for producing publicationready figures. Let's assume we have made a plot with some code (denoted by . . . below) containing all the information and data we want; we can then apply a theme like so:

```
. . . +
  theme_bw ()
```

In this example, we use + with the theme_bw function to use the built-in 'black and white' theme. This removes the grey background that some people dislike.

To find some of the available ready-made themes, type theme_ in the Console and hit the tab key. One popular alternative to theme_bw is the 'classic' theme theme_classic, which is also used in Figure 4.13. This produces a very stripped-down plot that's much closer to those produced by the base graphics system. Notice that we did one more thing there: we set the base_size argument to 7 to decrease the size of all the text in the figure (the default is 11). This can be used with any theme_XX function to quickly change the relative size of all the text in a plot.

7.4 Summing up and looking forward

We hope that clarifies how ggplot works. It took each of us quite some time to figure it out. Keep trying if it's not clear. The best way to get better is to practice and to use ggplot even for simple figures you might normally create in other ways. And don't let people who have been using R for years and 'don't get' ggplot put you off; we promise it's worth the extra effort at the start!

Among the four of us, we have used many graphing applications (e.g. Sigmaplot, Excel) and many of the different approaches in R. We have all now converted to ggplot, using it for everything from quick and dirty graphs that show us what we need to know (but that shouldn't be let out into the wild) to polished graphs for our articles, presentations, and reports. We are curious to know if, for the remainder of our careers working with data, we will use anything but ggplot. We can't imagine doing so, but do concede that it wasn't so long ago that we were not imagining using ggplot. Owen offers special thanks to Nick Isaac for mentioning, some years ago, that he should probably check out ggplot!

The **ggplot2** package has many abilities that we have not shown. More or less anything on a graph can be changed, i.e. colours, angle of text, size of text, and so on. You can even add GIFs and emojis! Google 'ggplot2 gallery' to get inspiration; you will probably then soon find the R Graph Gallery website.[3]

That is just about all of the R that we're going to expose you to in *Insights*. Well done. If you've followed our advice, you will now have a long list of functions and copious notes about what they do and how they work. As a reward for reaching this milestone, run this script:

```
install.packages("praise")
library(praise)
praise()
```

The next two chapters (which are the last before the conclusion chapter) concern some data analysis concepts, such as what variables are and what types of variables there are; populations, samples, and distributions; independence; exploring individual variables; and exploring relationships among multiple variables. But before then, as always, take a moment to review, consolidate, and relax. You've earned it!

---

[3] https://www.r-graph-gallery.com

Making deeper insights part 1
Working with single variables

Chapters 3 and 4 provided a rapid-fire introduction to many aspects of our workflow for acquiring insights from data. It highlighted many features of the process of importing and 'cleaning up' data to make it tidy and work well with functions for managing and manipulating data in the tidyverse. It also demonstrated ways to make some pretty effective figures that are 100% aligned with specific questions that were established at the outset of the analysis.

We have, however, done this with a minimal knowledge of the properties of the different variables, such as whether they are numeric or categorical, or of features of their distribution. In this and the next chapter, we'll delve deeper into these concepts that link data to appropriate visualizations and summaries that then lead to insights. This will also help you understand better how to refine your questions in advance of gathering the data.

In this chapter we will explore several conceptual issues raised by the bat diet Workflow Demonstration captured in Chapters 3 and 4. We will develop your knowledge of the following topics:

Insights from data with R: An Introduction for the Life and Environmental Sciences. Owen L. Petchey, Andrew P. Beckerman, Natalie Cooper and Dylan Z. Childs, Oxford University Press (2021). © Owen L. Petchey, Andrew P. Beckerman, Natalie Cooper and Dylan Z. Childs. DOI: 10.1093/oso/9780198849810.003.0008

- types of variables and collecting samples;
- exploring a numeric variable;
- exploring a categorical variable;
- dealing with missing data.

Before we do, we wish to emphasize two things we've mentioned many times before: the importance of carefully defining and recording your question and appropriately designing your approach (e.g. is it a survey, a controlled experiment, or an intervention?), and the core characteristics of data that influence how they can be analysed and what inferences can then be drawn (i.e. what types of questions we can answer).

Paying attention to these requires that we always step back to the basics. It's a good idea to remind ourselves frequently of:

- the number and names of the variables (columns) we are working with;
- the number of observations (rows) we've made;
- if and how many of the variables describe experimental manipulations.

Armed with these basics, let us revisit some of the intricacies of the bat diet dataset.

8.1 Variables and data

You will often be able to say something about the *kinds* of data you've collected before you even look at them. You'll probably know that some of the data are numeric, and some are not. Those that are not, you'll probably think of as categorical or factors. These are useful concepts. They help you make decisions about the types of graphs you can make and how and whether you can summarize the data into tables.

Here we want to dive a bit deeper into how variables can be classified. Statisticians have thought long and hard about how we should classify variables. The distinctions can be subtle but important for making appropriate insights. This is super-important for your choices about visualizing and summarizing data, and ultimately impacts on how you will build statistical models in the future. The classification makes us think hard about how observations vary among groups or along gradients.

We introduce two key classification schemes.

8.1.1 NUMERIC VERSUS CATEGORICAL VARIABLES

This is a classification scheme many of you will be familiar with. While acceptable, there are nuances that we think you should be aware of that can really help translate the interface between your questions and the data you've collected into robust insight.

Numeric variables have values that describe a measurable quantity as a number, like 'how many' or 'how much'. Numeric variables are also called quantitative variables; data collected containing numeric variables are called quantitative data. We can go further, though. Numeric variables may be further described as either continuous or discrete:

- **Continuous numeric variables**. Observations can take any value within certain set of real numbers, i.e. numbers represented by decimals. This set is typically either 'all the positive numbers' (e.g. biomass may be very large or very small, but it is strictly positive) or 'every possible number' (e.g. a *change* in biomass can be positive or negative). Examples of continuous variables include body mass, age, time, and temperature. Although, in theory, continuous variables may admit any number in the set of possible numbers, in practice the values given to an observation may be bounded and can only include values as small as the measurement protocol allows. Elephants are very large, but they never get as big as a large building; and trying to measure the mass of an elephant to a precision of a few grams is probably not practical.
- **Discrete numeric variables**. Observations can take a value based on a count from a set of whole values, e.g. 1, 2, 3, 4, 5, and so on. Sometimes

these are called 'integer variables'. A discrete numeric variable cannot take a value between one value and the next closest value. Examples of discrete variables include the number of individuals in a population, the number of offspring produced, and the number of individuals infected in an experiment. All of these are measured as whole units—there are no half-babies produced by organisms. Discrete variables are very common in the biological and environmental sciences. We love to count things…

Categorical variables have values that describe a characteristic of a data unit, like 'what type' or 'which category'. Categorical variables fall into mutually exclusive (in one category or in another) and exhaustive (include all possible options) categories. Therefore, categorical variables are qualitative variables and are most often represented by a non-numeric value—e.g. a word. In fact, it's a bad idea to represent categorical variables by numbers. The data collected for a categorical variable are qualitative data. Categorical variables may be further described as ordinal or nominal:

- **Ordinal variables**. Observations can take values that can be logically ordered or ranked. Examples of ordinal categorical variables include academic grades (i.e. A, B, C), size classes of a plant (i.e. small, medium, large), and measures of aggressive behaviour. The categories associated with ordinal variables can be ranked as higher or lower or earlier or later than one another, but do not necessarily establish a meaningful numeric difference between categories.
- **Nominal variables**. Observations can take values that cannot be organized in a logical sequence. Examples of nominal categorical variables include sex, business type, eye colour, habitat, religion, and brand. There is no order to the 'levels' of these categories.

One of the reasons that this distinction between ordinal and nominal is important is because you are likely want to ensure that the ordering of ordinal variables is represented in your visualizations and tables. You want to be able to ensure that readers of your work grasp this order quickly and align the patterns in the data with answers to the questions.

> *Don't use numbers to describe categorical information.* There is little to stop someone using numbers to describe a categorical variable (e.g. 1 = Terrestrial, 2 = Aquatic, 3 = Flying). But it usually isn't very sensible to do this. It's clearer to use the words themselves (we don't have to remember the coding) and helps when we make tables and graphs in R, as R will then use those names. We used to have sticky notes stuck to our monitors with codes… we not only forgot the codings all the time… we lost the sticky notes. Just use the words!

8.1.2 RATIO VERSUS INTERVAL SCALES

There is a second way of classifying numeric variables (*not* categorical variables), which relates to the scale they are measured on: ratio versus interval. The distinction is important because it determines whether certain types of calculations or comparisons are meaningful. Fundamentally, the difference between scales hinges on whether there is a 'true zero value'. Let's see what that means.

The distinction between ratio and interval scales is a property of the units, or what we call the *scale of measurement*. For example, it is the units of °C versus K (kelvin) that determine whether temperature is interval or ratio in a dataset. When we measure temperature in °C, we're working on an interval scale defined relative to the freezing and boiling temperatures of water under standard conditions. It doesn't make any sense to say that 30 °C is twice as hot as 15 °C. °C is interval. However, if we measure the same two temperatures on the kelvin scale, it is meaningful to say that 303.2 K is 1.05 times hotter than 288.2 K. This is because the kelvin scale is relative to a true zero: absolute zero. It is thus a ratio scale.

- **Ratio scale**. This scale has a simple idea: with ratio scale data, it is meaningful to say that something is 'twice as ...' as something else. This type of data possesses a meaningful zero value. It takes its name from the fact that a measurement on this scale represents a ratio between a measure of the magnitude of a quantity and a unit of the same kind. Ratio scales most often appear when we work with physical quantities. For example, we can say that one tree is twice as tall as another, or that one elephant has twice the mass of another, because length and mass are always measured on ratio scales.

- **Interval scale**. In contrast, this kind of scale does not have a unique and non-arbitrary zero value. However, interval data still allow for the degree of difference between data items, just not the ratio between them. A good example of an interval scale is date, which we measure relative to an arbitrary epoch (e.g. AD). We cannot say that 2000 AD is twice as long as 1000 AD. However, we can compare the amount of time that has passed between pairs of dates, i.e. their *interval*. For example, it's perfectly reasonable to say that twice as much time has passed since the epoch in 2000 AD versus 1000 AD.

Why should you care about this distinction? You may end up with poor insights if you don't. For example, when making a graph it doesn't make sense to automatically start your axis at 0 if your data are interval scale, whereas it often does if something is ratio. This matters because people tend to start making statements about ratios—e.g. a value is 'twice as big' as another—when you start the axis at 0.

8.2 Samples and distributions

When we collect data about some objects (e.g. animals, locations, cells), we are working with a sample of those objects. When we talk about 'exploring a variable', what we are really doing is exploring the **sample distribution** of that variable. What is this?

The sample distribution is a statement about the frequency with which different values occur in a particular sample. Imagine we took a sample of undergraduates and measured their heights. The majority of students would be round about 1.7 metres tall, even though there would obviously be some variation among students. Men would tend to be slightly taller than women, and very small or very tall people would be rare. We know from experience that no one in this sample would be over 3 metres tall. These are all statements about a (hypothetical) sample distribution of undergraduate heights.

The problem with samples is that they are 'noisy'. If we were to repeat the same data collection protocol more than once, we would expect to end up with a different sample each time. This results purely from chance variation and the fact that we can almost never sample everything we care about. Imagine we again went out, caught bats, and sampled their poop in order to collect a dataset like that in the Workflow Demonstration. We would never catch all the bats, we'd catch different bats each time, and even if we did happen to catch the same bat more than once (and know we had), we would get a different poop from that bat each time.

There are two fundamental insights to make about the sample distribution of a variable. The first is 'What are the most common values of the variable?' and the second is 'How much do the observations differ from one another?' As you know from Chapters 3 and 4, there are two ways to go about this:

1. *Calculate numerical summaries.* These are used to quantify the basic features of a sample distribution. They provide simple summaries about the sample that can be used to make comparisons and draw preliminary conclusions. For example, we often use 'the mean' to summarize the 'most likely' values of a variable in a sample. Some of you may be familiar with the use of variances or standard deviations to represent the variation in the data. Numerical summaries such as the sample mean are also called descriptive statistics.

2. *Construct graphical summaries.* Graphical summaries are an ideal complement to numerical summaries because they allow us to present lots of information about a sample distribution in one place and in a manner that is easy for people to understand. Moreover, visualizations help readers see both the question and the insight, as the axes, colours, fills, and points on a graph are organized to highlight questions and answers.

In the following sections we will develop deeper insights into single numeric and categorical variables. This will help reinforce your new-found understanding of these types of variables and provide the template on which we use this understanding with more complex visualizations in the next chapter.

8.2.1 UNDERSTANDING NUMERICAL VARIABLES

We'll work here with the `Wingspan_mm` and `No._Reads` (number of reads in the molecular analysis) variables in the bat diet dataset `bats` to illustrate the key ideas. Wingspan and number of reads are clearly numeric variables.

We can say a bit more. They are both numeric variables that are measured on a ratio scale, because zero really is zero: it makes sense to say that 20 mm is twice as large as 10 mm and 1000 reads is 10 times less than 10,000 reads.

But are these continuous or discrete numeric variables? Think about the possible values that each could take. A wingspan of 22.3425 is perfectly reasonable, so this could be a continuous numeric variable. It seems less reasonable to get 500.5 reads, however; whole numbers for reads seem more reasonable, so this is discrete/counts.

Look at the data to confirm what type of data are in each variable. Try the `glimpse` function:

```
glimpse(bats)
```

```
## Rows: 633
## Columns: 15
## $ Row_order   <dbl> 1, 2, 3, 4, 5, 6, 7, 8, 9, 10, 11, 12,
## $ Bat_ID      <dbl> 366, 366, 367, 367, 367, 367, 367, 367,
## $ Age         <chr> "Adult", "Adult", "Adult", "Adult",
## $ Sex         <chr> "Female", "Female", "Female", "Female",
## $ Sp._Nr.     <dbl> 52, 22, 64, 98, 114, 19, 2, 4, 70, 81,
## $ Species     <chr> "Ethmia bipunctella", "Bradycellus
## $ Class       <chr> "Insecta", "Insecta", "Insecta",
## $ Order       <chr> "Lepidoptera", "Coleoptera",
## $ Family      <chr> "Depressariidae", "Carabidae",
## $ Pest        <chr> "no", "no", "no", "no", "yes", "yes",
## $ Migratory   <chr> "no", "no", "no", "yes", "yes", "yes",
## $ Wingspan_mm <dbl> 23.5, NA, 31.0, 52.5, 40.0, 37.5, NA,
## $ No._Reads   <dbl> 32672, 411, 14796, 8953, 3472, 3282,
## $ Date        <chr> "20.07.12", "20.07.12", "20.07.12",
## $ Date_proper <date> 2012-07-20, 2012-07-20, 2012-07-20,
```

We see that the wingspan variable indeed contains numbers and some of these are fractions, but only to the nearest 0.5 mm, and all the visible values of the number of reads are integers. What if the researchers had measured wingspan to the nearest millimetre? Would this mean the wingspan was a discrete numeric variable? The underlying quantity is certainly continuous, but might 'look' discrete as a consequence of the way it was measured. This example illustrates an important idea: we shouldn't simply look at the values in a sample to determine what kind of variable we're working with. We have to think!

How we measure and categorize a variable is often a research decision. But we have to make this decision based on knowledge of its true nature and our measurement process. Just because 'real' wingspan is numeric and continuous, that does not necessarily mean it will be represented that way in a dataset. For example, the researchers might have used small, medium, and large as size categories, which would then make it an ordinal categorical variable.

Sometimes these distinctions just aren't clear. What if we were only able to measure wingspan to the nearest 5 mm instead of 0.5 mm? In this situation, we would definitely 'see' categories of wingspan and we could treat it as an ordinal categorical variable. However, it might also be reasonable to treat it as a numeric variable that just happens to have been measured on a very coarse-grained scale.

8.3 Graphical summaries of numeric variables

As you might recall from Chapter 4, we can use a histogram to create a visualization of the wingspan data (Figure 8.1):

```
ggplot() +
  geom_histogram(data = bats,
                 mapping = aes(x = Wingspan_mm),
                 bins = 10)
```

```
## Warning: Removed 78 rows containing non-finite values
## (stat_bin).
```

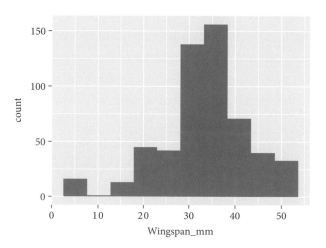

Figure 8.1 A histogram of the wingspan variable with 10 bins.

A central feature of histograms is bins—each bar in this histogram represents the count of the number of values of `Wingspan_mm` that fall within the *bin* of that bar. We discussed in Chapter 4 how binning works to help summarize lots of data.

Binning involves two steps. First, we take the set of possible values of our numeric variable and divide this into equal-sized, non-overlapping intervals. We can use any interval size we like, as long as it is large enough to span multiple observations some of the time, though in practice some choices are more sensible than others. We then have to work out how many times our variable falls inside each bin. The resulting set of counts tells us a lot about the sample distribution.

Here is one way to do the first step for the `Wingspan_mm` variable using the **ggplot2** `cut_interval` function:

```
# 10 bins example
bins <- bats %>%
    # grab the variable as a vector of numbers
    # (cut_interval does not take a data frame)
    pull(Wingspan_mm) %>%
    # create (10) intervals
    cut_interval(10)

# look at the first 10 entries of this
head(bins)
```

```
## [1] (20.3,24.9] <NA>      (29.5,34.1] (47.9,52.5] (38.7,43.3] (34.1,38.7]
## 10 Levels: [6.5,11.1] (11.1,15.7] (15.7,20.3] (20.3,24.9] ... (47.9,52.5]
```

The `cut_interval` function has created a set of categories to label our bins (`bins`) and assigned one of these to each wingspan measurement. We can see the specific bins using the `levels` function:

```
bins %>% levels()
```

```
## [1] "[6.5,11.1]"  "(11.1,15.7]" "(15.7,20.3]" "(20.3,24.9]" "(24.9,29.5]"
## [6] "(29.5,34.1]" "(34.1,38.7]" "(38.7,43.3]" "(43.3,47.9]" "(47.9,52.5]"
```

The object `bins` contains, as we wanted, *ten* bins. Each of these is labelled with its interval, defined by its beginning and end.

The second step is to find how many of the observed wingspans fall in each of the intervals/bins. We'll use a new function, `table`, from base R to do the counting of observations in each bin:

```
bins %>% table()
```

```
## .
##  [6.5,11.1]  (11.1,15.7]  (15.7,20.3]  (20.3,24.9]
##          16            8           22           33
## (24.9,29.5]  (29.5,34.1]
##          75          107
## (34.1,38.7]  (38.7,43.3]  (43.3,47.9]  (47.9,52.5]
##         152           69           21           52
```

These numbers are the heights that define the histogram bars. Take a look at the histogram in Figure 8.1. What does this tell us? It shows that most wingspan observations are round about 39 mm (in the middle of the [37.2, 42.3) bin). Values less than 16 mm are quite rare by comparison.

8.3.1 MAKING SOME INSIGHTS ABOUT WINGSPAN

These binned data tell us quite a lot about the sample distribution of `Wingspan_mm`. This is what a **histogram** shows us. The histogram in Figure 8.1 gives a nice summary of the sample distribution of the `Wingspan_mm` variable. It reveals (1) the most common values, which are around 39 mm; (2) the range of the data, which is about 50 mm; and (3) the shape of the distribution, which is asymmetric, with a tendency for the left 'tail' to extend further than the right 'tail'.

It's a good idea to experiment with the number of bins to see how the shape of the histogram varies as we do this. We set the properties of `geom_histogram` to tweak this kind of thing. Let's construct the histogram again with eight bins, as well as adjust the colour scheme and axis labels a bit (the resulting graph is in Figure 8.2):

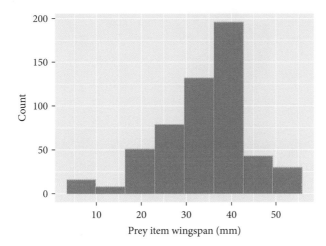

Figure 8.2 A histogram, as in the previous figure, but with eight bins and some other customization.

```
ggplot() +
  geom_histogram(data = bats,
                 mapping = aes(x = Wingspan_mm),
                 bins = 8,
                 fill = "steelblue",
                 colour = "darkgrey",
                 alpha = 0.8) +
  xlab("Prey item wingspan (mm)") +
  ylab("Count")
```

Whether or not that colour scheme is an improvement is a matter of taste! Mostly, we wanted to demonstrate that changing the number of bins can change the detail we perceive in the histogram.

We can use pretty much the same R code to produce a histogram summarizing the `number of reads` sample distribution (the resulting graph is in Figure 8.3):

```
ggplot() +
  geom_histogram(data = bats,
                 aes(x = No._Reads),
                 bins = 10,
                 fill = "steelblue",
                 colour = "darkgrey",
                 alpha = 0.8) +
```

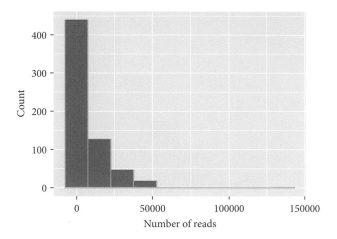

Figure 8.3 A histogram of the number of reads.

```
xlab("Number of reads") +
ylab("Count")
```

This histogram reveals that the No._Reads (number of reads) variable contains many small numbers and a few very large numbers: the shape of the distribution is asymmetric. The centre of the distribution is rather unclear, given its very asymmetric shape.

 Always remember when looking at histograms that we have made rather arbitrary/subjective decisions about the number of bins/bin width. It's good practice to check to make sure our impression of the shape of the distribution is not sensitive to the number of bins we choose.

We'll finish up this section by briefly reviewing two alternatives to the histogram. Histograms are good for visualizing sample distributions when we have a reasonable sample size (at least dozens, and ideally hundreds of observations). They aren't very effective when the sample is quite small. Nor are they the most effective graph for thousands or more of observations.

In the 'small data' situation, it's better to use something called a **dot plot**.[1] Let's use **dplyr** and the `sample_n` function to extract a small(ish) subset of the bat diet data:

```
bats_small <- bats %>%
  sample_n(100)
```

This samples 100 rows at random from the dataset. The **ggplot2** code to make a dot plot with these data is very similar to the histogram case (the resulting graph is in Figure 8.4):

```
ggplot() +
  geom_dotplot(data = bats_small,
               mapping = aes(x = Wingspan_mm),
               binwidth = 2) +
  xlab("Wingspan (mm)") +
  ylab("Count")
```

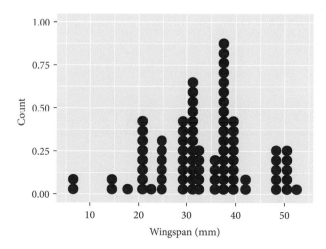

Figure 8.4 A dot-plot-type histogram can be useful for showing the sampling distribution of a small number of data points.

---

[1] Not to be confused with the 'Cleveland dot plot'. A standard dot plot summarizes a sample distribution. The Cleveland dot plot is something quite different; it summarizes the frequencies of a categorical variable. It's meant to serve as a simple alternative to bar charts and pie charts.

Here, each observation in the data adds one dot, and dots that fall into the same bin are stacked up on top of one another. The resulting plot displays the same information about a sample distribution as a histogram, but it tends to be more informative when there are relatively few observations. This is because the bins don't have to be regularly spaced in a dot plot. Instead, they are chosen in a way that attempts to best represent the distribution we're trying to summarize.

In the 'large data' situation, it is sensible to use a **density plot** (Figure 8.5). We can implement this on the raw data, as we did with the histogram:

```
ggplot() +
  geom_density(data = bats,
               mapping = aes(x = Wingspan_mm)) +
  xlab("Wingspan (mm)") +
  ylab("Count")
```

```
## Warning: Removed 78 rows containing non-finite values
## (stat_density).
```

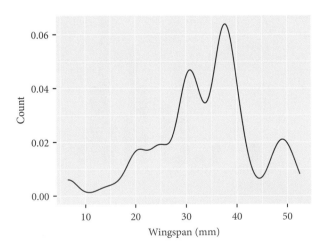

Figure 8.5 This is a density plot of the wingspan variable. Compare it with the histograms we made above.

8.3.2 DESCRIPTIVE STATISTICS FOR NUMERIC VARIABLES

So far, we've been describing the properties of sample distributions by examining graphs, using phrases like 'most common values' and 'the range of the data' without really saying what we mean. There are, however, quite specific terms that can be used to describe these kinds of properties. The two that matter most are the **central tendency** and the **dispersion**:

- A measure of **central tendency** describes a typical ('central') value of a distribution. Most people know at least one measure of central tendency. The 'average' that they calculated at school is the arithmetic mean of a sample. There are many different measures of central tendency, each with its own pros and cons. Among these, the mean is probably most often used, but often the median is a better option in exploratory analyses (we will see why below).
- A measure of **dispersion** describes how spread out a distribution is. Dispersion measures quantify the variability or scatter of a variable. If one distribution is more dispersed than another, this means that in some sense it encompasses a wider range of values. What this means in practice depends on the kind of measure we're working with. Basic statistics courses tend to focus on the variance and its square root, the standard deviation. However, the interquartile range is another one that is used often in exploratory analyses (we will see why below).

In the following two sections, we are going to develop an argument for the use of the median and interquartile range as effective measures of central tendency and dispersion during this exploratory phase of insight gathering. You've probably guessed by now, too, that we are going to pair these with some graphics. And you're right—we'll introduce the use of the box-and-whisker plot to display these after we review these measures.

8.3.3 MEASURING CENTRAL TENDENCY

There are two descriptive statistics that are typically used to describe the central tendency of the sample distribution of numeric variables. The first is the **arithmetic mean** of a sample. People often say 'sample mean' or even just 'the mean' when referring to the arithmetic sample mean. This is fine, but keep in mind that there are other kinds of mean (e.g. the harmonic and geometric means).

One limitation of the arithmetic mean is that it is quite affected by the shape (i.e. symmetry) of a distribution. This is why, for example, it does not make much sense to look at the mean income of workers in a country to get a sense of what a 'typical' person earns. Income distribution is highly asymmetric, and those few who are lucky enough to earn very good salaries tend to shift the mean upward and well past anything that is really 'typical'. The sample mean is also strongly affected by the presence of 'outliers'. (It's difficult to give a precise definition of outliers—the appropriate definition depends on the context—but, roughly speaking, these are unusually large or small values.)

Because the sample mean is sensitive to the shape of a distribution and the presence of outliers, we often prefer a second measure of central tendency: the **sample median**. The median of a sample is the value separating the upper half from the lower half of the distribution.

We can find the sample mean and sample median in R with the mean and median functions (no surprises there):

```
bats %>%
  summarise(mean_wingspan = mean(Wingspan_mm, na.rm = TRUE),
            median_wingspan = median(Wingspan_mm, na.rm = TRUE))
```

```
## # A tibble: 1 x 2
##   mean_wingspan median_wingspan
##           <dbl>           <dbl>
## 1          33.6              35
```

The mean and median of the wingspan are 33.6 mm and 35 mm, respectively. Notice that these seem to lie to the right of the most common values of the wingspan, but the sample median has been 'pulled' leftward less than the mean. That's what we meant when we said the mean is sensitive to the shape of a distribution.

8.3.4 MEASURING DISPERSION

Dispersion describes how 'spread out' the values of a sample are. The most important measures of dispersion from the standpoint of statistics are the **variance** and **standard deviation**. Variances and standard deviations are strictly non-negative. Smaller values indicate that observations tend to be close to one another, while high values indicate that observations are very spread out. Zero only occurs when all values are identical.

We can calculate the variance and standard deviation of a sample using the `var` and `sd` functions in R (you can guess which calculation each one does). However, we don't advise using these metrics for gathering insights about your data at the exploratory phase. Why do we say this?

- It is hard to make sense of variances unless you have a lot of experience of working with them. The calculation of a sample variance involves squaring the deviation of each data point from the overall sample mean. This aspect of the calculation means that it very difficult to assess whether the difference between two variances is 'big' or 'small'. It is also difficult to relate the perceived spread of distributions to their variance.
- Standard deviation is the square root of variance. This makes it a better statistic for describing sample dispersion, in the sense that it works on the same measurement scale as the variable you are summarizing. This makes it a bit easier to relate standard deviations to the spread of values we perceive when we graph a distribution. However, just as with the sample mean, standard deviations are sensitive to outliers.

Standard deviation and variance are important quantities in statistics that crop up over and over again. For example, many common statistical tools use changes in variance to formally compare how well different models describe a dataset. For understanding your data, however, it is generally easier to use a measure of dispersion that does not suffer from the problems we just outlined. The **interquartile range** is one such measure.

The interquartile range (IQR) is simply the range that contains 'the middle 50%' of a sample, which is given by the difference between the third and first quartiles. Take a look at the next information box if you haven't come across these before. The more spread out the data are, the larger the IQR. The reason we prefer to use the IQR to measure dispersion is that it is easy to interpret and, because it does not depend on the whole sample, it isn't affected so much by the presence of outliers.

We can use the `IQR` function to find the interquartile range of the wingspan variable:

```
bats %>%
  summarise(IQR_wingspan = IQR(Wingspan_mm, na.rm = TRUE))
```

```
## # A tibble: 1 x 1
##    IQR_wingspan
##           <dbl>
## 1            10
```

The IQR appears in a useful summary plot called a 'box-and-whisker' plot (see Figure 8.6). It is straightforward to calculate quartiles with **dplyr** functions:

```
bats %>%
  summarise(
    # labels for the quantiles
    quantile = scales::percent(c(0.25, 0.5, 0.75)),
    # estimate the quantiles
    q_wing = quantile(Wingspan_mm, c(0.25, 0.5, 0.75),
    na.rm = TRUE))
```

```
## # A tibble: 3 x 2
##    quantile q_wing
##    <chr>     <dbl>
## 1 25%          29
```

```
## 2 50%              35
## 3 75%              39
```

The IQR is hidden in this result. Remember, it is just the difference between the third and first quartiles $(39 - 29 = 10)$.

What are quartiles? We need to know what a quartile is to understand the interquartile range. Three quartiles are defined for any sample. These divide the data into four equal-sized groups, from the set of smallest numbers up to the set of largest numbers. The second quartile (Q_2) is the median, i.e. it divides the data into an upper and a lower half. The first quartile (Q_1) is the number that divides the lower 50% of values into two equal-sized groups. The third quartile (Q_3) is the number that divides the upper 50% of values into two equal-sized groups. The quartiles also have other names. The first quartile is sometimes called the lower quartile or the 25th percentile; the second quartile (the median) is the 50th percentile; and the third quartile is also called the upper quartile or the 75th percentile.

8.3.5 MAPPING MEASURES OF CENTRAL TENDENCY AND DISPERSION TO A FIGURE

Up to this point, you've learned about types of variables, reacquainted yourself more deeply with the histogram, and come to the new understanding that the median and IQR are 'robust' measures of central tendency and dispersion.

In addition to having the facility to calculate these on your own, it is also beneficial to generate a graphical representation of these measures (Figure 8.6). Not surprisingly, this is pretty straightforward:

```
ggplot()+
  geom_boxplot(data = bats,
               mapping = aes(x = "", y = Wingspan_mm),
```

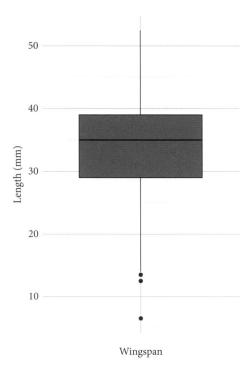

Figure 8.6 A boxplot of the wingspan variable showing the median, IQR, and various other features of the data distribution.

```
                fill = "hotpink") +
  xlab("Wingspan") +
  ylab("Length (mm)") +
  theme_minimal()
```

```
## Warning: Removed 78 rows containing non-finite values
## (stat_boxplot).
```

Notice that we forced **ggplot2** to hide the tick mark label on the *x*-axis by setting up an empty dummy label using x = " " in the mapping. We had to do this because **ggplot2** isn't really designed to plot a single boxplot. We also used theme_minimal() to remove graphical elements like axes, which aren't really needed for a simple plot like this.

The help file for geom_boxplot is pretty valuable for understanding how we link the above calculations of the median and IQR to the figure:

The boxplot compactly displays the distribution of a continuous variable. It visualises five summary statistics (the median, two hinges and two whiskers), and all 'outlying' points individually.... The lower and upper hinges correspond to the first and third quartiles (the 25th and 75th percentiles)... The upper whisker extends from the hinge to the largest value no further than 1.5 * IQR from the hinge (where IQR is the inter-quartile range, or distance between the first and third quartiles). The lower whisker extends from the hinge to the smallest value at most 1.5 * IQR of the hinge. Data beyond the end of the whiskers are called 'outlying' points and are plotted individually.

We now have several compact representations of the `Wingspan_` (mm) data, including the histogram, the boxplot, and the calculation of the IQR. You can generate the same summaries for the `No._Reads` data. Go for it! These tables and graphs provide the kinds of detail you need to understand the central tendency and dispersion of numeric variables. These are central, as we'll see, to comparing groups of data and drawing insights from these comparisons.

8.3.6 COMBINING HISTOGRAMS AND BOXPLOTS

Just for fun, we provide here some code to combine a histogram and a box plot for the `Wingspan_mm` data using a special package called **patchwork** (Figure 8.7):

```
library(patchwork) # you might need to install this first!

# make the histogram
# set the x-axis boundaries to 0:60
plt_hist <- ggplot(bats, aes(x = Wingspan_mm)) +
  geom_histogram(bins = 10, fill = "steelblue",
              colour = "darkgrey", alpha = 0.8) +
  xlim(0,60) +
  xlab("Wingspan_mm") + ylab("Count")

# make the boxplot
```

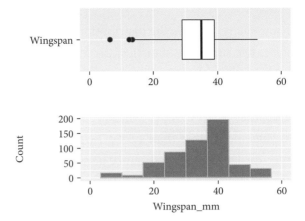

Figure 8.7 This figure and associated code demonstrate the use of the **patchwork** package to combine two **ggplot2** plots on the same page.

```
# set the y-axis boundaries to 0:60 to align with the histogram
# get rid of the axis labels
# coord-flip-it!
plt_box <- ggplot(bats, aes(x = "Wingspan", y = Wingspan_mm))+
  geom_boxplot()+
  ylim(0,60) +
  ylab("") + xlab("") +
  coord_flip()

# use patchwork syntax to place the boxplot on top of the
# histogram
# see also the cowplot and the gridExtra packages....
plt_box/plt_hist
```

8.4   A moment with missing values in numeric variables (NAs)

We introduced some discussion of missing values in Chapter 3. We'd like to come back to this because it is very important to be intimately aware of the number of data points we're working with, and where there are NAs in our data. This can influence the estimation and number of estimates returned for summary measures of central tendency and dispersion.

We saw that the bat diet dataset contains 78 missing values of the `Wingspan_mm` variable, and no missing values in any other variables. We then calculated three response variables for each bat:

```
prey_stats <- bats %>%
  group_by(Bat_ID, Sex, Age) %>%
  summarise(num_prey = n(),
            num_prey1 = n_distinct(Sp._Nr.),
            mean_wingspan = mean(Wingspan_mm, na.rm = TRUE),
            prop_migratory = sum(Migratory == "yes") / n())
```

```
## `summarise()` regrouping output by 'Bat_ID', 'Sex'
    (override with `.groups` argument)
```

A few things here are worth noting:

- We might have decided to remove all the rows in the data frame containing NAs in the `Wingspan_mm` variable before doing anything else. But then we would have lost some information from the dataset on variables that did not have any NAs—potentially useful and important information. Also, we might have created some bias if there was something special about the NAs.
- The NAs are all for non-Lepidoptera prey, so any analysis of wingspan will only be complete for Lepidoptera only. We needed to know this. (Check this by viewing the data.)
- One bat (`Bat_ID  1320`) ate only one prey item, and this was non-Lepidoptera (check this again by viewing the data or using the `filter` function). Hence any measure of central tendency and dispersion we calculate will result in one fewer estimate than we expect, because only 142 bats have non-NA mean-wingspan entries. This also means that we will have 143 estimates of the other response variables that we summarize by `Bat_ID`.

R functions will usually take care of NAs. They will then often give a warning like '5 rows with missing values removed'. If you see such a warning, absolutely, without a doubt, 100% make sure you know why these are being removed, and put a note in your code. To be clear, if you let functions take care of NAs then make sure you know what the functions are doing. And it's probably better to take care of them yourself, explicitly.

8.5 Exploring a categorical variable

Phew! That's a lot to take in. Have you had your biscuit brain food?

We are now going to consider how to go about exploring the sample distribution of a categorical variable. Just to refresh your memory, don't forget that there are ordinal and nominal categorical variables. Check back at the start of this chapter for why this distinction might be valuable!

8.5.1 UNDERSTANDING CATEGORICAL VARIABLES

Exploring categorical variables is generally simpler than working with numeric variables because we have fewer options for estimating central tendencies and dispersion. We'll work with the `Family` and `Migratory` variables in `bats` to illustrate the key ideas.

We can use the `distinct` function to identify the unique categories and gain some initial feeling for them:

```
bats %>%
  distinct(Family)
```

```
## # A tibble: 25 x 1
##    Family
##    <chr>
## 1 Depressariidae
## 2 Carabidae
## 3 Noctuidae
```

```
## 4 Pyralidae
## 5 Crambidae
## # ... with 20 more rows
```

```
bats %>%
  distinct(Migratory)
```

```
## # A tibble: 2 x 1
##    Migratory
##    <chr>
## 1 no
## 2 yes
```

A first question we can ask is, is `Family` an ordinal or nominal categorical variable? There is no particular way to order these, so it's a nominal variable. What about the `Migratory` variable? There are only two values, yes or no, so we are treating this as a nominal variable too. This lack of ordering means we can proceed directly to making summaries.

Numerical summaries

When we make summaries of categorical variables we are still aiming to describe the sample distribution of the variable, just as with numeric variables. However, the general question we need to address is 'what are the relative frequencies of different categories?' This is related to what we did with using histograms and bins to assess the distribution of numeric variables above.

To do this, we need to identify which categories are common and which are rare. Since a categorical variable takes a finite number of possible values, the simplest thing to do is count the number of occurrences of each type. The **dplyr** `count` function does this super-quickly for us and allows us to sort the presentation from high to low:

```
# the 'print' bit at the end lets us see it all
bats %>% count(Family, sort = TRUE) %>% print(n = nrow(bats))
```

```
## # A tibble: 25 x 2
##    Family                                        n
##    <chr>                                     <int>
##  1 Noctuidae                                   371
##  2 Geometridae                                  75
##  3 Crambidae                                    28
##  4 unclassified Lepidoptera                     22
##  5 Tortricidae                                  20
##  6 Gelechiidae                                  18
##  7 Tipulidae                                    15
##  8 Chrysopidae                                  14
##  9 Pyralidae                                    14
## 10 Depressariidae                               11
## 11 unclassified Arthropoda                       9
## 12 Lygaeidae                                     8
## 13 Erebidae                                      4
## 14 Hemerobiidae                                  4
## 15 Noctuidae / Geometridae / Tortricidae         3
## 16 Gelechiidae/ Geometridae                      2
## 17 Lycaenidae                                    2
## 18 Noctuidae/ Crambidae                          2
## 19 Noctuidae/ Gelechiidae/ Noctuidae             2
## 20 Nolidae                                       2
## 21 unclassified Diptera                          2
## 22 unclassified Insecta                          2
## 23 Carabidae                                     1
## 24 Culicidae                                     1
## 25 Curculionidae                                 1
```

This shows that the numbers of species per taxonomic family are very different, ranging from 371 for Noctuidae to one observation for some families (e.g. Carabidae). Raw frequencies such as these give us information about the rates of occurrence of different categories in a dataset. However, it's difficult to compare raw counts across different datasets if the sample sizes vary (which they usually do). This is why we often convert counts to proportions. To do this, we have to divide each count by the total count across all categories:

```
# get the proportion number of observations in the Family variable
prop_obs_per_fam <- bats %>%
  count(Family, sort = TRUE) %>%
  mutate(Prop_obs = n / sum(n))

# look at what we've made....
prop_obs_per_fam

# to see all 25 rows.... un-comment the next line!
# prop_obs_per_fam %>% print(n = nrow(prop_obs_per_fam))
```

```
## # A tibble: 25 x 3
##    Family                       n Prop_obs
##    <chr>                    <int>    <dbl>
## 1 Noctuidae                  371    0.586
## 2 Geometridae                 75    0.118
## 3 Crambidae                   28    0.0442
## 4 unclassified Lepidoptera    22    0.0348
## 5 Tortricidae                 20    0.0316
## # ... with 20 more rows
```

So about 59% of species are Noctuidae, with the next most observed being the Geometridae, at around 12% of observations. Can you find the set of least represented families using `tail()`?

What about measuring the central tendency of a categorical sample distribution? Basically, this isn't possible or sensible with categorical variables. We can certainly estimate the **sample mode** of ordinal and nominal variables easily: this is just the most common category. For example, we know from above that Noctuidae is the modal value of the `Family` variable. It is important to remember that this isn't really *central tendency*. In the world of categorical variables, it is *commonness*.

It is, however, possible to calculate a **sample median** of an ordinal categorical variable. Remember, the median value is the one that lies in the middle of an ordered set of values—it makes no sense to talk about 'the middle' of a set of nominal values that have no inherent order. Unfortunately, even for an ordinal variable the sample median is not precisely defined. Imagine that we're working with a variable with only two categories, 'big' versus 'small', and exactly 50% of the values are 'small' and 50% are large. What is the median in this case? Because the median is not always well defined, the developers of base R have chosen not to implement a function to calculate the median of an ordinal variable. A few packages contain functions to do this, though.

 Be careful with median. Unfortunately, if we apply the median function to a character vector it will give us an answer. It is very likely to give us the wrong answer, though. R has no way of knowing which categories are 'high' and which are 'low', so it just sorts the elements of type alphabetically and then finds the middle value of this vector. If we really have to find the median value of an ordinal variable, we can do it by first converting the categories to integers—assigning 1 to the lowest category, 2 to the next lowest, and so on—and then use the median function to find out which value is the median.

So where does that leave us? For numeric variables, we focus on summary estimates of *central tendency* and *dispersion*. But for categorical variables, these are not defined well, and we focus on counting, looking carefully at the counts and calculating proportions. However, in common with numeric variables, there is a graph we can make that ultimate really helps with gaining insights.

Graphical summaries of categorical variables

When summarizing a single categorical variable, the length of the bars can show the raw counts or proportions of observations in each category. We can get a quick view of the raw data with the following code that produces Figure 8.8:

```
ggplot() +
  geom_bar(data = bats,
           mapping = aes(x = Family))
```

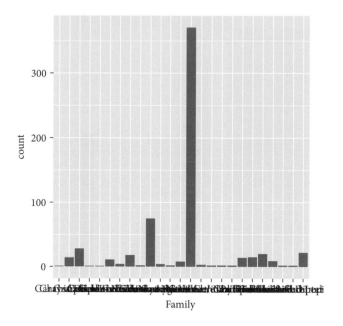

Figure 8.8 A bar plot of counts of the data occurring in a categorical variable, here counts of the number of values recorded for each family of prey. We will fix the overlapping x-axis tick labels later.

This plot is not too bad. We can see which families are common and which are rare, though it's quite hard to get a sense of their *relative* representation.

There are also a lot of different categories with long names, which means we've ended up with serious overplotting of the x-axis tick mark labels. We'll review one way to address that problem in a bit.

A different graphical tool for summarizing a categorical variable is a *stacked bar chart* (Figure 8.9). A stacked bar chart uses the lengths of bars stacked on top of one another to capture the proportion of observations in each category. Let's do this for the Family variable. We'll present two methods—one using the raw data and geom_bar and one using the proportions we actually calculated above with geom_col.

Here is the method that uses the raw data (data = bats) and geom_bar:

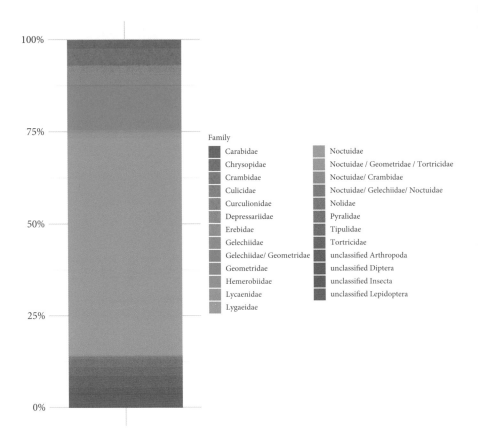

Figure 8.9 A stacked bar chart using the raw data.

```
# raw data method
ggplot(data = bats,
       aes(x = "", fill = Family)) +
  geom_bar(position = "fill") +
  scale_y_continuous(labels = scales::percent) +
  xlab("") + ylab("") +
  theme_minimal()
```

We specify `fill = Family` to ensure that the stacked bars are defined by different colours for different families, and use `geom_bar(position = "fill")` to force `geom_bar()` to calculate the proportions of cases in each family for us. Sweet and simple.

What is going on with the *x*-axis? We used `aes(x = "")` to set up an 'empty' dummy variable associated with the *x*-axis and `xlab("")` to suppress the labelling of the axis. The legend already tells readers that we're looking at family information, so there's really no need to add redundant information by labelling the *x*-axis.

What have we done with the *y*-axis? We added a prettification via `scale_y_continuous(labels = scales::percent)` to turn the *y*-axis tick labels into percentages. We also suppressed the main *y*-axis label because it's kind of obvious that we're looking at percentages, so there is no need to flag that by adding a label.

What do you think of this graph? It's OK, if not exactly great. The height of each bar represents the proportion of bat diets associated with each family. We can see a large central block associated with the Noctuidae, and lots of families with very small representations. You might also be able to tell that Geometridae is the next most common family represented in bat diets and pretty much every other family is relatively rare.

It's quite hard to pick out additional insights because there are so many categories, which means **ggplot2** uses a lot of similar colours to delineate each family.

Finally, here is the method that uses the `prop_obs_per_fam` summary dataset we calculated earlier and `geom bar`:

```
# with pre-calculated proportions
ggplot(data = prop_obs_per_fam,
      aes(x = "", y = Prop_obs, fill = Family))+
  geom_col() +
  xlab("")
```

We haven't bothered to show the graph this time. Trust us, it's essentially the same as before. This time we use our pre-calculated proportions (`y = Prop_obs`) from the `prop_obs_per_fam` data, and we now use `geom_col()` ('col'umn).

Notice that `geom_col()` is used when we've pre-calculated things, while `geom_bar()` is used when we want **ggplot2** to calculate counts for us. Both methods are a bit clunky in this example because we're making a *single* stacked bar plot. **ggplot2** wasn't really designed to do this easily. Usually, as you'll see in the next chapter, stacked bar plots are used to understand two or more categorical variables. We do this in Chapter 9 with migratory status and family.

The better visualization is probably the first graph we made, with separate bars for each family. We can customize this bar graph with functions like `xlab` and `ylab`, and by setting various properties inside `geom_bar`. We also had a bit of an issue with not being able to see the names of the families on the *x*-axis—the categories of `Family` have quite long names, meaning the axis labels are all bunched together. One way to fix this is to just flip the *x*- and *y*-axes to make a horizontal bar chart. We can do this with the `coord_flip` function, which makes sufficient space for the names, and chooses a good font size:

```
ggplot() +
  geom_bar(data = bats,
           mapping = aes(x = Family),
           fill = "steelblue",
           width = 0.7) +
  xlab("Family") +
  ylab("Number of Observations") +
  coord_flip()
```

The resulting graph is shown in Figure 8.10. This is the same as Figure 8.8 but with the coordinate system flipped. Put another way, with the *x*- and *y*-axes flipped.

8.6 Summing up and looking forward

The content of this chapter is essential for exploring data to gain insights. We have focused on single variables to delve into how we summarize and visualize different data types. You've learned about the idea of distributions, how to estimate the central tendency and dispersion of numeric and categorical variables, and how to visualize these metrics—these are the foundations of exploratory and statistical analyses.

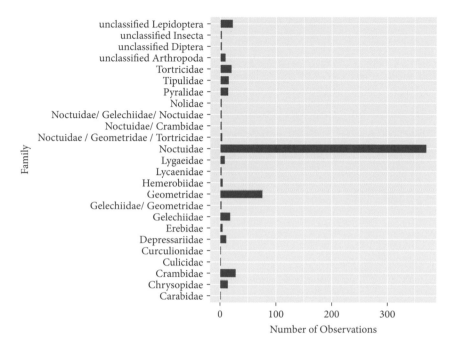

Figure 8.10 The original counts for each family, flipped to see the labels.

In the next chapter you will move to developing tools for understanding relationships among variables, like how `Wingspan_mm` varies with `Sex` or `Age`. Looking at and understanding relationships among variables is one of the most important tasks in getting insights from data.

In preparation for this, have a look back at the definitions of the different types of variables at the beginning of this chapter and review some of the details in the Workflow Demonstration in Chapters 3 and 4, in particular those about answering our questions (i.e. section 4.4: this is where we first started to look at relationships among variables in the workflow).

8.7 A cat-related reward

It is a bit tricky to install the package and get it working, so for now just have a look at this website showing how one can make various cat-themed and related graphs in R.[2]

---

[2] https://github.com/Gibbsdavidl/CatterPlots/blob/master/README.md

Making deeper insights part 2

Relationships among (many) variables

In the previous chapter we looked at individual variables, making numeric and graphical summaries to reveal the central tendencies and features of their sample distributions. Looking at and understanding relationships among variables is one of the most fundamental tools we use to align the questions we ask with the data we collect to gain insights.

This is because questions often boil down to understanding the relationships among two or more variables. These relationships might involve the same type (e.g. numeric versus numeric) or different types (e.g. numeric versus categorical) of variable. The main goal of this chapter is to show how to use descriptive statistics and visualizations to explore associations among different kinds of variables: two numeric variables, two categorical variables, and finally one numeric and one categorical variable.

Whether we are working with numeric or categorical variables, we are aiming for insight about how the values of one variable depend on those of the other. **Associations** between *pairs* of variables in a sample are called **bivariate associations**. An association is any relationship between two variables that makes them dependent, i.e. knowing the value of one variable gives us some information about the possible values of the second variable.

Insights from data with R: An Introduction for the Life and Environmental Sciences. Owen L. Petchey, Andrew P. Beckerman, Natalie Cooper and Dylan Z. Childs, Oxford University Press (2021). © Owen L. Petchey, Andrew P. Beckerman, Natalie Cooper and Dylan Z. Childs. DOI: 10.1093/oso/9780198849810.003.0009

So, if you have not done so already, now is a good time to refresh your memory about the types of numeric and categorical variables; their properties guide the type of insights you can make from summaries and visualizations of two or more variables.

One other thing to note here too. In this chapter we are going to introduce graphical summaries and descriptive statistics. This is because we have a strong philosophy, when we teach, of making a picture. The picture always aligns with a question and gives you a satisfying view of whether or not what you expected to see has materialized in the data you collected. Sometimes we also need a number we can report—descriptive statistics— to summarize the pattern we observe.

9.1 Associations between two numeric variables

9.1.1 DESCRIPTIVE STATISTICS: CORRELATIONS

Statisticians have devised various different ways to quantify an association between numeric variables. The common measures seek to calculate some kind of **correlation coefficient**. The terms 'association' and 'correlation' are closely related, so much so that they are often used interchangeably. Strictly speaking, correlation has a narrower definition: a correlation is defined by a metric (the 'correlation coefficient') that quantifies the degree to which an association tends to a certain pattern.

There are several measures of correlation, some of which you may have come across: the **Pearson's correlation coefficient**, which is good for describing linear associations, and then **Kendall's** and **Spearman's rank correlation coefficients**, which are useful when the assumptions for Pearson's correlation coefficient are not met (see below).

Pearson's correlation coefficient (also called the Pearson product-moment correlation coefficient) is designed to summarize the strength of a *linear* (i.e. 'straight line') association. Pearson's correlation coefficient takes a value between −1 and 1. A value of 0 arises when there is no linear association between two variables. A value of +1 or −1 arises w they are perfectly related. 'Perfectly related' means we can predict the exact value of one variable given knowledge of the other. The sign and the magnitude are valuable insights. A positive value indicates that high values of one variable are associated with high values of the second. A negative value indicates that high values of one variable are associated with low values of the second. The closer the correlation is to −1 or +1, the more closely related the variables are.

In R, we can use the `cor` function to calculate Pearson's correlation coefficient. Let's use the `prey_stats` dataset from Chapter 4, which contain the response variables `prey_num` and `mean_wingspan`, and calculate the Pearson correlation coefficient between them.

A few things to demystify. The `cor` function is not designed to work with data frames. There is no argument `data =`. Therefore, we need to make sure `cor` knows where the data come from. An easy way to do this with functions that lack this data argument is to use the `with` function, which is what we've done here:

```
prey_stats %>%
  filter(!is.nan(mean_wingspan)) %>%
  with(cor(num_prey,mean_wingspan, method = "pearson"))
```

```
## [1] -0.17
```

The result −0.17. What insights have we gained from this? Remember that there are two features of this coefficient—its sign and its magnitude. The coefficient is negative, indicating that bats eating more prey types tend also to eat smaller prey. The value −0.17 is also closer to 0 than to −1, suggesting that the variables are not so closely related.

Pearson's correlation coefficient must be interpreted with care. Two points are worth noting:

1. Because it is designed to summarize the strength of a *linear* relationship, Pearson's correlation coefficient will be misleading when this relationship is curved or, even worse, hump-shaped. Advice: plot the data first!

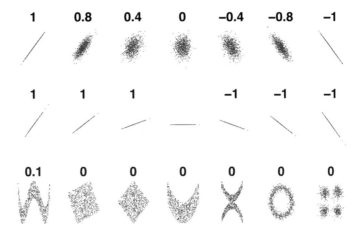

Figure 9.1 A variety of different relationships between pairs of numeric variables. The numbers in each subplot are the Pearson's correlation coefficients associated with the patterns.

Code for figure adapted from Denis Boigelot's code on Wikimedia Commons.[1]

2. Even if the relationship between two variables really is linear, Pearson's correlation coefficient tells us nothing about the slope or gradient (i.e. the steepness) of the relationship. Remember that the slope describes how a change in one variable determines the change in another.

If those last two statements don't make immediate sense, take a close look at Figure 9.1. Consider each row of the figure:

1. The first row shows a series of linear relationships that vary in their strength and direction. These are all linear in the sense that the general form of the relationship can be described by a straight line. This means that it is appropriate to use Pearson's correlation coefficient in these cases to quantify the strength of association, i.e. the coefficient is a reliable measure of association.

---

[1] https://commons.wikimedia.org/wiki/File:Correlation_examples2.svg

2. The second row shows a series of linear relationships that vary in their direction, but are all examples of a perfect relationship— we can predict the exact value of one variable given knowledge of the other. What these plots show is that Pearson's correlation coefficient measures the strength of association without telling us anything about the steepness of the relationship.

3. The third row shows a series of different cases where it is definitely inappropriate to Pearson's correlation coefficient. In each case, the variables are related to one another in some way, yet the correlation coefficient is always 0. Pearson's correlation coefficient completely fails to indicate the relationship, because it is not even close to being linear.

9.1.2 OTHER MEASURES OF CORRELATION

What should we do if we think the relationship between two variables is non-linear? We should not use Pearson's correlation coefficient to measure association in this case. Instead, we can calculate something called a **rank correlation**. The idea is quite simple. Instead of working with the actual values of each variable, we 'rank' them, i.e. we sort each variable from lowest to highest and then assign the labels 'first', 'second', 'third', etc. to different observations. Measures of rank correlation are based on a comparison of the resulting ranks. The two most popular are **Spearman's** ρ ('rho') and **Kendall's** τ ('tau').

Both coefficients are interpreted in just the same way as Pearson's correlation coefficient, with a value between -1 and 1, and their sign and magnitude are useful for making insights.

We can calculate both rank correlation coefficients in R using the same R syntax, combining `with` and `cor`. But this time we need to set the `method` argument to the appropriate value: `method = "kendall"` or `method = "spearman"`. For example, the Spearman's ρ and Kendall's τ measures of correlation between `num_prey` and `mean_wingspan` are given by

```
with(prey_stats, cor(num_prey, mean_wingspan, use = "complete.obs",
  method = "kendall"))
```

```
## [1] -0.1559375
```

```
with(prey_stats, cor(num_prey, mean_wingspan, use = "complete.obs",
  method = "spearman"))
```

```
## [1] -0.2306176
```

These roughly agree with Pearson's correlation coefficient, though Kendall's τ seems to suggest that the relationship is weaker. Kendall's τ is often smaller than Spearman's ρ. Although Spearman's ρ is used more widely, it is more sensitive to errors and discrepancies in the data than Kendall's τ.

9.1.3 GRAPHICAL SUMMARIES BETWEEN TWO NUMERIC VARIABLES: THE SCATTERPLOT

Correlation coefficients give us a simple way to summarize associations between numeric variables. They are limited, though, because a single number can never summarize every aspect of the relationship between two variables. This is why we always visualize their relationship. We suggest, in fact, that you never calculate a correlation coefficient without also plotting the data.

The standard graph for displaying associations among two numeric variables is at scatterplot, using horizontal and vertical axes to plot two variables as a series of points. We saw how to construct scatterplots using **ggplot2** and, in particular, the use of `geom_point` in Chapter 7, so we won't step through the details again.

There are several ways to enhance a basic scatterplot to not only reveal the nature of the relationship, but also reveal features of the distribution of each variable, as was introduced in Chapter 8.

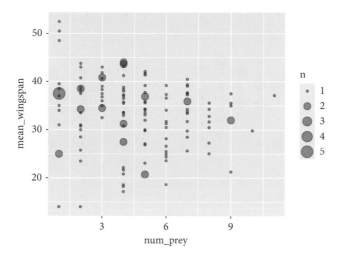

Figure 9.2 A scatterplot with the point size mapped to the number of data points lying on top of each other.

ggplot2 provides a couple of different geom_XX functions for producing a visual summary of relationships between numeric variables that can highlight the frequency of different values. This is also valuable for situations where overplotting of points is obscuring the relationship.

One such geom is the geom_count function (with the resulting graph in Figure 9.2):

```
ggplot() +
  geom_count(data = prey_stats,
            mapping = aes(x = num_prey, y = mean_wingspan),
            alpha = 0.5)
```

The geom_count function is used to construct a layer in which data are first grouped into sets of identical observations. The number of cases in each group is counted, and this number ('n') is used to scale the size of the points. It's a bit like creating bins, like we learned with respect to histograms.

Take note—it may be necessary to round numeric variables first (e.g. via mutate) to make a usable plot if they aren't already discrete.

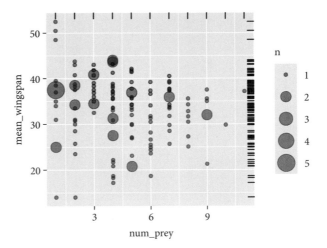

Figure 9.3 We have remade the scatterplot shown in Figure 9.2, but now included a mini-histogram called a rug on each axis.

We can add even more information about the distribution of points by using a rug plot with the scatterplot (Figure 9.3). geom_rug is a handy geom for adding what is essentially an unbinned histogram on each axis:

```
ggplot(data = prey_stats,
       mapping = aes(x = num_prey, y = mean_wingspan)) +
  geom_count(alpha = 0.5) +
  geom_rug(sides = "tr")
```

```
## Warning: Removed 1 rows containing non-finite values
## (stat_sum).
```

What we have here in this figure now is a scatterplot representing the correlation between the two variables, axes with (warm and cosy) rugs on them detailing the distribution of data for each variable, and point sizes representing the frequency of observations at each value. This single figure contains many of the details we discussed about single variables as well as the new bivariate detail. Notice that rugs don't work so well with a discrete variable like num_prey, because overplotting in the rug means we can't really tell if particular values are common or not.

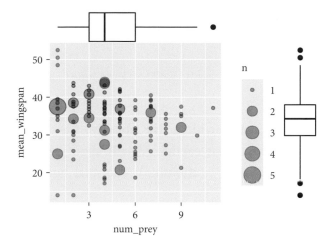

Figure 9.4 Using the **ggExtra** package, we can add marginal histograms to the scatterplot, and rugs for a very comprehensive graphical summary.

Two further options for dealing with overplotting of data points are the geom_bin_2d and geom_hex functions, the latter in the **hexbin** package. You could also combine geom_point with a large point size and apply alpha transparency, allowing the colour saturation to reflect the intensity of data at each position.

Finally, the **ggExtra** and **cowplot** packages (remember these?) allow us to place histograms on the x and y margins of the graph. A simple Google search on 'histogram, margins, ggplot' will lead to joy and happiness and hours of diversion/procrastination. Here is a quick look at what **ggExtra** can do via its ggMarginal function (Figure 9.4):

```
library(ggExtra) # you need to install this first

# make the base plot
p.base <- ggplot(data = prey_stats,
                 mapping = aes(x = num_prey, y = mean_wingspan)) +
  geom_count(alpha = 0.5)

# add the base plot to ggMarginal and request marginal plots
# these can be "boxplots" or "histograms"!
ggMarginal(p.base, type = "boxplot")
```

```
## Warning: Removed 1 rows containing non-finite values
## (stat_sum).

## Warning: Removed 1 rows containing non-finite values
## (stat_sum).
```

These efforts are not worthless. The objective has been to devise a graphical representation that captures details about both variables and their correlation (bivariate relationship). The final figure we produced generates:

- graphical insight into the central tendency of each variable;
- graphical insight into the spread of data in each variable;
- graphical insight into the frequency of data across the range of the data;
- graphical description of the weak negative correlation between the two variables.

These are the kinds of visualization that you should be aiming to make!

9.2 Associations between two categorical variables

9.2.1 NUMERICAL SUMMARIES

Numerically exploring associations between pairs of categorical variables is not as simple as the numeric-variable case. The general question we need to address is, 'Do different *combinations* of categories seem to be under- or over-represented?' We need to understand which combinations are common and which are rare. The simplest thing we can do is 'cross-tabulate' the number of occurrences of each combination. The resulting table is called a **contingency table**. The counts in the table are sometimes referred to as frequencies. If we do this effectively, we'll identify how many observations are associated with each combination of values of Family and Migratory.

The xtabs function (= 'cross-tabulation') can do this for us. For example, the frequencies of each combination of Family and Migratory is given by

```
xtabs(~ Family + Migratory, data = bats)
```

```
##                                    Migratory
## Family                             no yes
##    Carabidae                        1   0
##    Chrysopidae                     14   0
##    Crambidae                        8  20
##    Culicidae                        1   0
##    Curculionidae                    1   0
##    Depressariidae                  11   0
##    Erebidae                         3   1
##    Gelechiidae                     18   0
##    Gelechiidae/ Geometridae         2   0
##    Geometridae                     36  39
##    Hemerobiidae                     4   0
##    Lycaenidae                       2   0
##    Lygaeidae                        8   0
##    Noctuidae                      175 196
##    Noctuidae / Geometridae / Tortricidae  0   3
##    Noctuidae/ Crambidae             0   2
##    Noctuidae/ Gelechiidae/ Noctuidae  2   0
##    Nolidae                          2   0
##    Pyralidae                       14   0
##    Tipulidae                       15   0
##    Tortricidae                     20   0
##    unclassified Arthropoda          9   0
##    unclassified Diptera             2   0
##    unclassified Insecta             2   0
##    unclassified Lepidoptera        22   0
```

The first argument sets the variables to be cross-tabulated. The xtabs function uses R's special formula language, so we can't leave out that ~ at the beginning. After that, we just provide the list of variables to be cross-tabulated, separated by a + sign. The second argument tells the function which dataset to use. This isn't a **dplyr** function, so the first argument is *not* the data for once.

What does this tell us? It shows us how many observations are associated with each combination of values of `Family` and `Migratory`. We have to stare at the numbers for a while, but eventually it should be apparent that migratory prey are relatively infrequent, though a few `Family` entries do contain more migratory than non-migratory species.

If both variables are ordinal, we can also calculate a descriptive statistic of association from a contingency table—a table containing the frequency of combinations of each *level* among the categorical variables; this is a table that can be produced, for example, by the `xtabs` function. To do this we have to use some kind of rank correlation coefficient that accounts for the categorical nature of the data. One measure of association that is appropriate for such cross-tabulated, ordinal, categorical data is Goodman and Kruskal's γ ('gamma'). This takes a value of 0 if the categories are uncorrelated, and a value of +1 or −1 if they are perfectly associated. The sign tells us about the direction of the association. Unfortunately, there isn't a base R function to compute Goodman and Kruskal's γ, so we have to use a function from one of the packages that implements it (e.g. the `GKgamma` function in the **vcdExtra** package) if we need it.

9.2.2 GRAPHICAL SUMMARIES

Bar charts can be used to summarize the relationship between two categorical variables. The basic idea is to produce a separate bar for *each combination* of categories in the two variables. The lengths of these bars are proportional to the values they represent, which are usually either the raw counts or the proportions in each category combination. This is the same information as that displayed in a contingency table. You will likely recall that we did this in the second part of the bat Workflow Demonstration (Section 4.5).

Using **ggplot2** to display this information is not very different from producing a bar graph to summarize a single categorical variable.

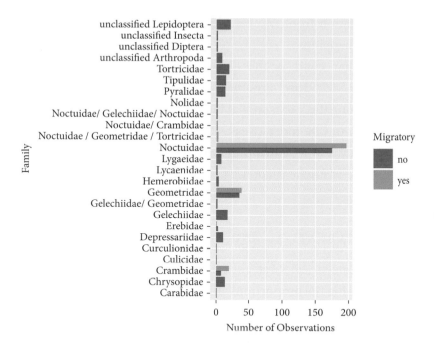

Figure 9.5 A stacked bar plot showing the number of prey items for each family, for each type of prey, migratory or not migratory

Let's do this for the `Family` and `Migratory` variables in `bats`, with the resulting graph shown in Figure 9.5:

```
ggplot() +
  geom_bar(data = bats,
           mapping = aes(x = Family, fill = Migratory),
           position = "dodge") +
  coord_flip() +
  labs(x = "Family",
       y = "Number of Observations")
```

A family is associated with two bars when it contains both migratory and non-migratory species in the poop samples, and only one bar otherwise. The length of each bar shows the number of observations associated with a combination of family and migratory status. Notice that we've included two aesthetic mappings to make this happen. We mapped the `Family` variable to the *x*-axis and the `Migratory` variable to the fill colour. We want to display information from two categorical variables, so we have to define two aesthetic mappings.

A stacked bar chart is the default produced by `geom_bar`. The `position = "dodge"` argument says that we want the bars to 'dodge' one another along the *x*-axis so that they are displayed next to one another. We also snuck in one more tweak. Remember, we can use `labs` to set the labels of any aesthetic mapping we've defined—we used it here to set the label of the aesthetic mapping associated with the fill colour and the *x*/*y*-axes.

9.2.3 AN ALTERNATIVE, AND PERHAPS MORE VALUABLE

Insights from categorical × categorical associations are actually more easily made via graphing the proportions. The valuable plot is a stacked bar chart of the proportions of migratory and non-migratory across the families.

We can make a plot very easily by asking `geom_bar` to link `fill = Migratory` to `position = "fill"`, which forces the calculation of the proportions of migratory and non-migratory species for each family (Figure 9.6). Clever clogs, this `geom_bar`—as we saw in Chapter 8, it is doing some work for us:

```
ggplot(bats, aes(x = Family, fill = Migratory)) +
  geom_bar(position = "fill") +
  scale_y_continuous(labels = scales::percent) +
  coord_flip()
```

The figure shows that among families of insects in the bat diets, most have 0% migratory species. There are a couple of families with only migratory species and a couple of families with a mixture. This is a much more accessible figure than that of the raw counts presented above in Figure 9.5, particularly when at least one of the categorical variables (e.g. `Family`) has more than a few levels. However, we do lose one bit of useful information in this plot: sample size. That is, we can't tell how many species went into the proportions represented in this figure.

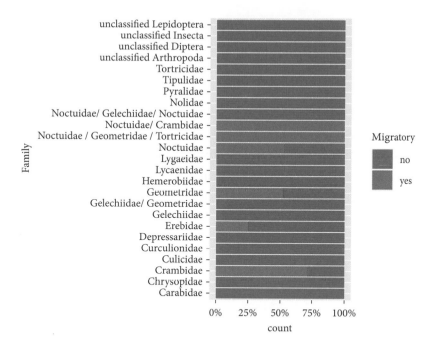

Figure 9.6 A stacked bar chart showing the migratory versus non-migratory counts in each family.

9.3 Categorical–numerical associations

We've seen how to summarize the relationship between a pair of variables, and gain insight from numerical and graphical summaries, when the variables are of the same type: numeric versus numeric or categorical versus categorical. The obvious next question is, 'How do we display the relationship between a categorical and a numeric variable?' This is one of the relationships that we are are most frequently required to visualize and summarize. The simplest case is associated with asking whether two groups differ from each other! We do this all the time.

As usual, there are a range of different options. Recall that we have already done this several times when we were answering our three questions (one for each response variable) in the bat Workflow Demonstration in Section 4.4.4. But let's look a bit deeper and connect the key ideas around central tendency and dispersion from Chapter 8 to this two-or-more-groups problem.

9.3.1 NUMERICAL SUMMARIES

Numerical summaries can be constructed by taking the various ideas we've explored for numeric variables (means, medians, etc.), and applying them to subsets of data defined by the values of the categorical variable. For example, we can calculate the mean wingspan (the numeric variable) of the prey of male and female (the categorical variable) bats. This is straightforward to do with the **dplyr** `group_by` and `summarise` pipeline, as you've seen throughout the previous chapters (see Section 5.2.2). While these tabular summaries are helpful, they don't provide the array of insights a graphical summary can. Let's see how to use these types of summary data graphically. As you've seen in the Workflow Demonstration (Chapters 3 and 4) and Chapter 8, this revolves around the boxplot and histogram.

9.3.2 GRAPHICAL SUMMARIES FOR NUMERICAL VERSUS CATEGORICAL DATA

In the bat Workflow Demonstration, we used `geom_beeswarm` to show the relationship between a numeric response variable and a categorical explanatory variable. Showing the data like this can be a nice way to view the raw data but it can make it difficult to extract important features of distributions. This is particularly true if we're lucky enough to have a lot of data to look at.

An alternative method builds on the box-and-whisker plot ('boxplot') we introduced in the previous chapter. As we noted there, this plot summarizes some important features of numeric variables' distributions. A boxplot summarizes the median, interquartile range, and more extreme values of a numeric variable. Compared with quantities like the mean and standard deviation, these tend to be less affected by the odd extreme value, or outlier, when present.

Here we extend the use of the boxplot to compare two groups. Let's start by plotting the proportion of migratory prey for male versus female

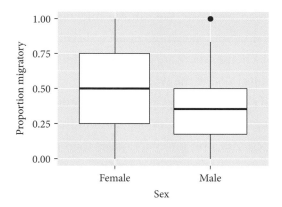

Figure 9.7 A box-and-whisker plot summarizing the distribution of the proportion of migratory prey species for each bat sex.

bats. To do this, we use the `prey_stats` dataset we made in Chapter 4 again.

Recall that, in Chapter 8, we used `geom_boxplot(x = "Wingspan_mm", y = Wingspan_mm)` to force the geom to label the *x*-axis with the single variable we were summarizing. Here we are asking for two boxplots to be produced, one each for bat sex. We do this by specifying the bat sex variable for the *x*-axis and our proportion for the *y*-axis: `geom_boxplot(data = prey_stats, mapping = aes(x = Sex, y = prop_migratory))`, with the resulting graph shown in Figure 9.7:

```
ggplot() +
  geom_boxplot(data = prey_stats,
               mapping = aes(x = Sex, y = prop_migratory))+
  xlab("Sex") +
  ylab("Proportion migratory")
```

We refer you to the help file for `geom_boxplot`, which is also reproduced in part in Chapter 8, for the definitions of the central line (median), boxes (interquartile range), and other lines and dots. The resulting plot compactly summarizes the distribution of the numeric variable within each of the categories.

What insight can we gain? We can see information about the central tendency, dispersion, and skewness of each distribution. In addition, we can get a sense of whether there are potential outliers by noting the presence of individual points outside the whiskers.

Moreover, and aligned with the question of whether the sexes differ in the proportion of migratory species among families in their diet, it shows that `Prop_migratory` is a bit lower for males than for females but the two distributions overlap quite a lot and are relatively symmetric in both sexes. Does this surprise you given that there are so few species that are migratory in the diets, as we saw above in the stacked bar chart of the proportions? Do you think there are enough data to actually make inferences about these small differences? Do you have an opinion now (an insight) about whether the sexes differ? These are the kinds of questions we get to ask ourselves by making these informative plots!

9.3.3 ALTERNATIVES TO BOX-AND-WHISKER PLOTS

Box-and-whisker plots are a good choice for exploring categorical–numerical relationships. They provide a lot of information about how the distribution of the numeric variable changes across categories. Sometimes we may want to squeeze even more information about these distributions into a plot.

One way to do this is to make multiple histograms, or dot plots if we don't have much data. We already know how to make a histogram, and we have seen how aesthetic properties such as `colour` and `fill` are used to distinguish different categories of a variable in a layer. This suggests that we can overlay more than one histogram on a single plot.

Let's use this idea to see how the sample distribution of wingspan (`Wingspan_mm`) differs among the two sexes (with the resulting graph in Figure 9.8):

```
ggplot()  +
  geom_histogram(data = prey_stats,
                 mapping = aes(x = mean_wingspan, fill = Sex),
```

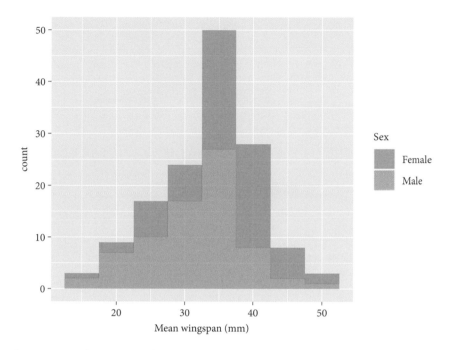

Figure 9.8 A histogram of the prey wingspan distribution for each sex of bat.

```
                 alpha = 0.6, binwidth = 5) +
   xlab("Mean wingspan (mm)")
```

We define two mappings: the continuous variable (Wingspan_mm) is mapped to the x-axis, and the categorical variable (Sex) is mapped to the fill colour. The default behaviour of geom_histogram is to overlay the two histograms, which is why we've specified alpha = 0.6 to allow transparency.

Plotting several histograms in one layer like this places a lot of information in one plot, but it can be hard to make sense of this when the histograms overlap a lot. As an alternative, we might consider producing a separate histogram for each category, and we've already seen a quick way to do this via *facets* (Section 7.1.4). Faceting works well here (with the resulting graph in Figure 9.9):

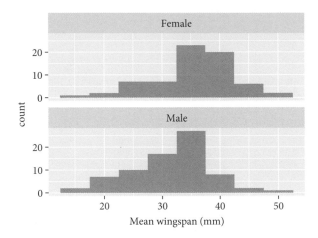

Figure 9.9 The same two histograms as in the previous figure, but this time each in its own facet.

```
ggplot()  +
  geom_histogram(data = prey_stats,
                 mapping = aes(x = mean_wingspan),
                 alpha = 0.8,
                 binwidth = 5) +
  xlab("Mean wingspan (mm)") +
  facet_wrap(~ Sex, ncol = 1)
```

Notice that we've specified ncol = 1. This is deliberate. When exploring data with histograms, it can be a good idea to stack them on top of each other so that estimates of central tendency and the distribution can be compared vertically. geom_histogram and facet_wrap work smartly together to ensure that the *x*-axis is the same for each facet/panel. This makes it easy for you and colleagues to start making insights and forming opinions about differences.

Super! And what do we conclude from this figure? It looks like female bats eat prey of greater wingspan than males, with the difference being around 5 mm. But there is also a lot of overlap in the distributions and not a great deal of difference in their dispersion, minimum, or maximum. (We saw this also in Figure 4.6.)

We suggest now that you use the histogram method with the proportion-migratory data and the boxplot method with these wingspan data. Get comfortable with both—they are each super-useful for making insights from your data aligned with core questions about whether there are differences between groups.

9.4 Building in complexity: Relationships among three or more variables

We examined various plots that summarize associations between two variables in the last section. How do we explore relationships between more than two variables in a single graph? That is, how do we explore **multivariate associations**? It's difficult to give a concrete answer to this question, because it depends on the question we're trying to address, the kinds of variables we're working with, and, to a large extent, our creativity and our aptitude with a powerful graphing framework like **ggplot2**. Nonetheless, we already know enough about how **ggplot2** works to build some fairly sophisticated visualizations. There are two ways to add additional information to a visualization:

1. Define aesthetic mappings to allow the properties of a layer to depend on the different values of one or more variable.
2. Use faceting to construct a multipanel plot according to the values of categorical variables.

We can adopt both of these approaches at the same time, meaning we can get information for four to six variables into a single graph if we need to (though this does not always produce an easy-to-read plot). We've already seen these two approaches used together in Chapter 7. We'll look at one more example to illustrate the approach again.

Say we want to understand how the sample distribution of mean wingspan varies over the course of a year. We also want to visualize how this differs between sexes. One way to do this is to produce histograms

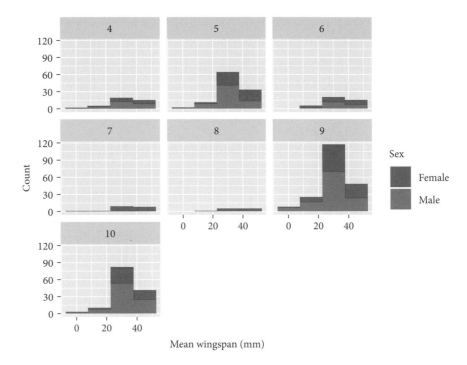

Figure 9.10 Histograms of prey wingspan for each sex of bat, with a facet for each month in which the data were collected.

for each month of the year, where the colour of the stacked histograms changes with respect to sex. We do this using the `facet_wrap` function to specify separate panels for each month, colouring the histograms by the `Sex` variable. Stacking the histograms happens by default (with the resulting graph in Figure 9.10):

```
ggplot() +
  geom_histogram(data = bats,
                 mapping = aes(x = Wingspan_mm, fill = Sex),
                 binwidth = 15) +
  xlab("Mean wingspan (mm)") +
  ylab("Count") +
  facet_wrap(~ month(Date_proper), ncol = 3)
```

Notice that we're using `bats`, the unsummarized raw dataset. We haven't used any new tricks here (except pulling the month out using the `month` function from the **lubridate** package). We just set a couple of aesthetics and used faceting to squeeze many histograms onto one plot. It shows that May, September, and October are pretty good times for bats. And, perhaps interestingly, that only female bats were caught in July and August.

Here is a fun challenge: try changing `geom_histogram` to `geom_boxplot`—this just works. Then try setting `ncol = 7` instead of `ncol = 3`. Or even try `ncol = 1`! You are learning key skills that will allow you to develop core insights with complex data structures. Again, we can see how the key graphical summaries, the histogram and boxplot, can communicate a wide range of data patterns.

Over the course of this chapter, we've built on core principles from Chapters 3, 4, and 8. We've seen how to align questions with numerical summaries and with graphical summaries. Critically, here we've seen how to capitalize on histograms, boxplots, scatterplots, point sizes, point transparency, and facets to deliver deeply insightful visualizations that capture complexity in data aligned with our core questions.

9.5 Summing up and looking forward

Discovering and examining relationships, and then attempting to understand their causes, is quite a large proportion of how people gain an understanding of how our world and universe work. But great caution is required. It is very easy to perceive randomness as pattern, and to detect patterns that have little practical significance (formal statistical analyses can help us here, to quantitatively estimate the weight of evidence). It also seems to be very easy to assign reasons (causes) to patterns without going through rigorous methods to show that a candidate cause is actually the cause. Manipulations (coupled with formal statistical analyses) are one quite reliable method for showing causes of patterns, and are present in two of our Workflow Demonstrations. In the other two, we would need further studies to test if hypothetical causes of patterns were in fact operating.

That is about it. All that remains is the final chapter of this book, for which you more or less have to only sit back and read.

Looking back and looking forward

We have covered a lot of ground, learning many of the fundamental skills associated with reliably, efficiently, and confidently getting insights from data. These skills range from how to use R and RStudio to understanding concepts such as distributions and associated graphical rep resentations. You are equipped with a general and generalizable workflow that can be used, revised, and modified depending on your questions and data! What next, then?

Our first advice is to perform a study of your own, from question to insight. It doesn't have to be anything groundbreaking. If you have already made such a study and have some data, but did not use R, then have a go at putting into practice some of your new skills. Few things are as powerful as using your new skills on your own projects.

But don't worry if you feel you have not yet mastered the skills, and don't worry if you forget some things. Google is a great place to search for the things we forget, especially for questions about how to get R to do what we want. There are also a lot of free tutorials out there that cover some of the same material as this book.

Insights from data with R: An Introduction for the Life and Environmental Sciences. Owen L. Petchey, Andrew P. Beckerman, Natalie Cooper and Dylan Z. Childs, Oxford University Press (2021). © Owen L. Petchey, Andrew P. Beckerman, Natalie Cooper and Dylan Z. Childs. DOI: 10.1093/oso/9780198849810.003.0010

Oh, and here's something to keep in mind: you will get frustrated with R. You'll forget commas and brackets. You'll misspell function names and wrongly capitalize variable names. The four of us now have a sixth sense for when we've done something stupid in R (it's ringing most of the time!) ... it helps us know when we've done something particularly silly, but unfortunately often doesn't help us know what we've done. Humour, particularly the ability to laugh at ourselves when we find that we wrote a pipe instead of a plus when using `ggplot`, is great therapy.

There is a lot we decided to leave out of the book that we would have loved to put in. So here are some pointers about what we left out, and what you could now start exploring. In time, we hope to put more information about each of these on the *Insights* companion website.[1]

10.1 Next learning steps

Code style. The phrase 'code style' refers to the rules that govern how we structure code, how we name things, and so on. We didn't much mention code style in this book, though we did attempt to use a consistent and not too unorthodox style throughout (even if you didn't notice us doing it). Unlike most other programming languages, R is kind of the wild west when it comes to code style conventions, and, unfortunately, there isn't a universally accepted standard for how to write 'good' R code. Still, it is important to be aware of what the elements of good code style are, and to decide what you will adopt and adhere to in your own work. Iegor Rudnytskyi's guide to R code style[2] is a solid starting point. You don't have to choose that one, but do try to identify a good, widely used code style you like ... and stick to it! This will make it much easier for you to write readable and maintainable scripts in the long run. Your current and future collaborators will also thank you for making the effort.

---

[1] http://insightsfromdata.io
[2] https://irudnyts.github.io//r-coding-style-guide/

Visual communication. We have introduced a workflow for getting data into R, wrangling them into shape, and then examining them in various ways to gain insights. That last step often involved making some kind of graph, i.e. a data visualization. What we have not done so much of in this book is discuss what makes a really effective professional data visualization. When you move from making quick informative plots to producing 'publication-ready' figures it is really important to think about every aspect of the visualization: the range of the axes, the size and type of fonts, the properties of the colour scheme, the text annotations, and so on. Paying close attention to these details takes effort, but doing so will help you communicate your work effectively to the widest possible audience. How do you learn to make good choices? There are a lot of good resources out there, but the one we like most is *Fundamentals of Data Visualization*[3] by Claus Wilke. You could do a lot worse than starting there (and it is free!).

Statistical analysis. We talked a fair bit about distributions in this book and presented you with several options to visualize these (e.g. histograms, dot plots, and boxplots). We did this because insights ultimately stem from understanding the variability in your data. What we did not discuss are ideas such as probability and uncertainty, and how *these* abstract concepts relate to the patterns we've been unpicking. We mentioned the normal, Poisson, and binomial distributions during the bat Workflow Demonstration, but we didn't tell you much about the processes that might generate data with these distributions. Learning about these ideas is necessary to transition from garnering informal insights to using formal methods to test our ideas. These more formal methods fall under the umbrella of 'statistics'. We have deliberately avoided introducing many statistical concepts in this book. The world is full of introductory statistics books, and we wanted to introduce you to the all-important steps that need to happen before you dive into statistical analysis. That said, you will eventually have to start using those tools if you want to convince others that your insights are 'real', rather than simply a triumph of hope over reality.

---

[3] https://clauswilke.com/dataviz/

10.2 Reproducibility: What, why, and how?

We mentioned in the Preface how science is rightly under increasing pressure to be more open and demonstrably reproducible. We are strong proponents of openness and reproducibility, so feel it appropriate to close this book with some information about that. It seems appropriate also because if we can instill openness and reproducibility early in people's learning about getting insights from data, there is the potential for them to become core features of people's practices and workflows.

A *reproducible* study is generally thought of as one for which an independent person can reproduce the conclusions (insights) from the data. That is, a reproduction would not usually involve collecting new data. The independent person may or may not be able to reproduce the insights from the data.

A *replication* study would collect new data, and then check if those data led to the same insights. Many high-profile studies have proven impossible to replicate, leading to confusion and the erosion of trust in experts. Our work should have less importance if it cannot be replicated, and may even cause doubts to arise.

Here we focus on increasing the chance that our work is *reproducible*, i.e. that someone else (or ourselves) could reproduce the insights derived from the data. There are many tools to help us achieve this. Because of the focus of this book, here we will mention tools for making the data exploration (and analysis) stages reproducible. Spoiler alert—one of those tools is the tidyverse in R, and we have already shown you how to use it in this book . . .

10.2.1 WHY SHOULD YOU TRY AND MAKE YOUR WORK REPRODUCIBLE?

Hopefully we all want to do good work, and, increasingly, funding bodies are asking to see the data and code used to run analyses. Additionally, the kinds of work we can do are getting more complicated as more data become available, and we have more ways of working with them. This also means more ways of making mistakes, or of subconsciously affecting the outcome of our analyses. Making your work reproducible often helps to prevent these.

When people talk about reproducibility we often assume we are making our work reproducible for someone else, but we often question ourselves much more than we question others. Having a reproducible record of our work, e.g. an R script, which contains all of our instructions, can be very reassuring. Not reassuring in the sense that it is necessarily correct. But just that we have it to refer to, to check, and to correct. For example, you might be changing a graph after your friends/colleagues/supervisor/boss made some recommendations, or might be dealing with reviewer comments after submitting a paper. This will be so much quicker if you think about reproducibility throughout your work. Trust us, it's worth it!

10.2.2 HOW CAN YOU MAKE YOUR WORK MORE REPRODUCIBLE?

Some basic ideas for making your insights reproducible are listed below. If you're interested in learning more, the British Ecological Society *Guide to Reproducible Code*[4] is a good place to read after what comes below.

1. *Organize your projects.* Getting organized is a good way to make sure you don't lose code or data, and that you're always using the right data for the right project. We have already recommended that you have a folder for each project/analysis, and within this have a folder called `data` to pop your data into. You might also want to have folders called `analyses`, `thesis`, `figures`, etc. to tidy the contents further as you progress. The more projects you're working on, the more important it is to be organized!

2. *Avoid setting the working directory. Use relative paths instead.* We mentioned this in the Workflow Demonstration, but it's worth reiterating. To read in data, etc., R needs to know where we have stored these data. In

---

[4] https://www.britishecologicalsociety.org/wp-content/uploads/2017/12/guide-to-reproducible-code.pdf

old scripts, you might see that people set this in R at the start of their script using `setwd("path-to-folder")`. This works fine until you move the folder, or change computers, or send the code to your friend/supervisor to take a look at. Then the code breaks because R can't find your data.

Instead, we recommend using *projects* in RStudio. If you make a project in the folder for your project, R automatically knows to look in that folder. If you move the folder, R still knows where to look. If you copy that folder to another computer, R still knows where to look. If you send that folder to a friend, R *still* knows where to look. Projects are magic.

If you have a project in a folder with some data, double-click on the project to open RStudio. You can then read in the data without telling R anything other than the name of the data file:

```
lovely <- read_csv("my-lovely-data.csv")
```

You'll notice that we have to do something a little different when we put our data into folders within the project folder. For example, if the data are in the `data` folder, we instead use

```
lovely <- read_csv("data/my-lovely-data.csv")
```

This tells R to look into the `data` folder. Using `data/` in this way is called using **relative paths**, i.e. the location is relative to the overall project folder. Likewise, if we want to save a figure into the `figures` folder we can use relative paths to do this:

```
ggsave("figures/my-fancy-plot.png")
```

3. *Do not edit your raw data 'by hand'.* In data analysis courses you'll often hear people say that 'the raw data are sacrosanct'. This means that you should carefully check your Excel files for errors and correct them after data entry. But after this you should make those files read-only and make modifications using only a script in R, not by editing the Excel or CSV file. Why? This allows you more flexibility in deciding how to deal with your

data. You're free to make changes and try out different things, all while keeping the raw data for you to go back to if you make a mistake.

For example, in many places in this book we have removed variables to make analyses easier, but we did so with the copy in R's brain, not in the original file. We wouldn't want to remove these permanently from the dataset—we might need them later! Having a script for all your data tidying and cleaning also means that if at some point the data are updated— perhaps you/your colleagues add a new experiment, go on another field trip, or discover some data sheets at the back of a filing cabinet that haven't been entered into the data file—you don't need to do all these modifications in Excel again; you can just quickly rerun your beautiful R script.

How do you make these edits? In R, of course, using some of the tricks and functions we've taught you in this book (for example `mutate`, `filter`, `rename`, `gather`, `spread`, etc.).

If you have to do a lot of data cleaning, or if you have lots of data files that you're going to stick together using `joins`, it can help to have a separate folder called `raw-data` and to keep all the raw data there. At the end of the script you can then use another handy **readr** function, `write_csv`, to make a new CSV file of the tidied and combined dataset for you to save in the `data` folder and to use in gaining your insights. We do something similar to this in the food-diversity–polity Workflow Demonstration in the online material.

4. *Use scripts to make insights (graphs, statistics, etc.).* Throughout this book we have been making our insights using an R script, so hopefully this point is already second nature to you by now! The advantage of this is we can save the script and rerun the code at any point we like in the future. Awesome!

Sometimes, for very complicated graphs for publication, you might need to edit them using other software, but we recommend avoiding this as much as possible. **ggplot2** is really flexible and the online help is immense, so you should be able to make most kinds of graph. For reproducibility purposes, this means you can quickly edit and recreate any graphs you need.

 It can sometimes be tempting to fiddle around with things in the Console and not to save them to a script—don't do this! Always work in a script. If it's code for just playing around with the data, you can always use the # to comment it out later, or if you're sure it isn't needed you can delete it. But always work with scripts.

5. *Think about your coding style and be consistent.* You'll have noticed in this book that there are lots of different ways to do things in R. For example, each of the following bits of code does exactly the same thing:

```
# 1
lovely%>%filter(evil==FALSE)%>%pull(nice)

# 2
lovely %>%
filter(evil==FALSE) %>%
pull(nice)

# 3
lovely %>%
  filter(evil == FALSE) %>%
  pull(nice)
```

Our preferred style is 3—it's tidy and it's clear what is going on. It is also what happens naturally and automatically when you use RStudio.

But how you code is a matter of choice, although we recommend that you never write code like 1, as it's very hard to read, especially when you have lots of steps. Tidy code is easier to read. Various *coding style guides* exist (here is one we like[5]), but the most important thing is to be consistent.

---

[5] http://adv-r.had.co.nz/Style.html

If you're happy to use RStudio's coding style, you can automatically tidy messy code by highlighting it, going to the Code menu, and then selecting Reformat Code. Easy. Peasy. Lemon. Squeezy.

6. *Write code and comments for humans, not computers.* If you've ever done any R or programming before, you'll be familiar with this scenario. You work on some code, write a few comments to explain to yourself what you were trying to do, save the code, and then do something else for a day/week/month/year. You then reopen the code and sigh in despair! What idiot wrote this?! What on earth was I thinking?!

There are two ways to help with this problem. First, write *human-readable code*. We recommend in this book that you use short names for objects, etc. so they are easy to type, but variables called X1, X2, and X3 will be nonsensical when you go back to the code later. Instead, try and find short names that are slightly more descriptive, for example head, tail, and foot. If you do need to use numbers or abbreviations, make sure to write thorough comments about what these mean.

Additionally, if you're doing something complicated, try and write it in a way that is easy for you as a human to read, even if it takes a few more steps. For example, the two bits of code below do the same thing, but the first is much harder to read.

```
# 1
pull(filter(lovely, evil == FALSE), nice)

# 2
lovely %>%
  filter(evil == FALSE) %>%
  pull(nice)
```

Another thing to remember is to write plenty of comments. Always write far more than you think you'll need. Write comments about *why* you are doing something. You don't always need to write comments on bits of code that are obvious; for example, in the code above, pull(nice) is clearly pulling out the entries of the variable nice, assuming we remember what the function pull does (and we can look this up quite quickly even if we don't). But why we might only want the values from `nice` is not clear, so we should make comments to remind ourselves why we were doing this.

7. *Share the data, metadata, and code.* A final, but very important, step towards reproducibility is to share the data and metadata. Metadata are data about your data(!). The header row of a spreadsheet is metadata . . . data about the data below. So are the explanations we provided of the variables in the Workflow Demonstrations. Many repositories exist for storing such data, including Dryad[6] and figshare,[7] and many institutions also have their own. There are many considerations when preparing your data for sharing that are beyond the scope of this book, so we recommend taking a look at the British Ecological Society Guide to Data Management[8] for more details.

Likewise, you should also be prepared to share your code. People are understandably nervous about doing this. What if people see my code and think I'm an idiot? What if people find mistakes in my code? Hopefully we can dispel the former worry—there is so much code online that very little of it is ever looked at. Also, everyone was a beginner once, and 99.9% of people are very accepting of that, and actually want to help you to improve, not to shoot you down (the other 0.1% of people are not worth engaging with).

Other people finding errors in your code is something a lot of people worry about, but really this is a positive thing. Yes, it's excruciatingly embarrassing at first, but better to discover an error than to work with incorrect insights. And again most people want to help, and to make the work better, not score points with your errors.

There are many places to share code. If you want to go further with this, you may want to set up a GitHub[9] or Bitbucket[10] account. You can share code on web pages. Or you could just add it as a supplemental file to a paper or thesis chapter or assignment. At the very least, make sure to save

---

[6] https://datadryad.org/
[7] https://figshare.com/
[8] https://www.britishecologicalsociety.org/wp-content/uploads/Publ_Data-Management-Booklet.pdf
[9] https://github.com/
[10] https://bitbucket.org/

the code and keep a version ready in case anyone needs it in the future, and also share it with your colleagues/supervisor/boss in case they need it too.

8. *Further stuff.* There are lots of other things you can do that are beyond the scope of this book, including using a version control system like **git**, writing reproducible reports using **RMarkdown**, unit testing, functional programming, etc. If you are interested in learning more, the British Ecological Society *Guide to Reproducible Code*[11] is a good place to start.

> Reproducibility can seem really scary, but before you panic or refuse to think about it, remember the following. (1) Even small modifications to how you work can make your insights *a lot* more reproducible. Don't feel you need to do all the things we suggest here; just choose one or two per project and you'll still reap the rewards. (2) You already know how to do many of the most helpful things (e.g. using projects and relative paths, and working with scripts to tidy/clean data and make graphs)—they're part of this book!

10.3 Congratulations!

Very well done if you worked through the whole book. Very well done if you are happy that you learned what you wanted to. Very well done if you tried hard but are not satisfied with your learning. Don't forget our advice to now do some work with your own data. This will solidify your experience with and knowledge of getting insights from data and make it more personal. And, finally, if you have any comments about the content of this book or the online material, especially if they may improve the content, please email us. Thanks!

---

[11] https://www.britishecologicalsociety.org/wp-content/uploads/2017/12/guide-to-reproducible-code.pdf

Index

Note: Tables and figures are indicated by an italic *t* and *f* following the page number.